Destiny

by Michael A. Torres Sr.

508 West 26th Street KEARNEY, NE 68848
402-819-3224
info@medialiteraryexcellence.com

Contents

Dedication

I dedicate this book to my soulmate of fifty years plus, of unconditional love, affection, and dedication, who has been and will continue to be my destiny from the Lord Almighty, to be with for the rest of my life, my loving wife, Elba Luz. I also dedicate this book to the rest of my family for their moral support.

Acknowledgments

I have to give thanks to my grandson, Christian Miliano Reyes, for his knowledgeable, technical, and immense support in helping me start and complete this project.

Introduction

May 4, 1966 I find myself in a Capital International Airways Jet. Our pilot has just announced that we will be stopping for fuel in Anchorage, Alaska, after eight hours of flying time. We will then proceed to Tokyo, Japan, for another refuel after another eight hours in the air. Finally after six more hours in the skies, we would arrive at our final destination. For somebody who hates to fly, twenty-two hours in the air is just a little bit too much for me. To say the least, I was nervous and scared at the same time. I wished it was all a dream but it was not. It was really happening to me.

There are hundreds of thoughts running through my mind during this long flight. I felt I was just sentenced to die even though I was only twenty years old. My mind took me back to my early childhood....

Chapter I

VIEQUES ISLAND

I was remembering my precious early childhood in the island of Vieques. It is a beautiful island belonging to Puerto Rico but located to the east of the mainland. Its size is approximately 21 miles long by 4 miles wide with about 10,000 inhabitants. The island is all breathtaking!

Born here, my island of paradise! The shore is lined up with tall palm trees being moved from side to side by the pleasant breeze. The aqua blue-green waters are so warm to the touch, the waves so calm as to be saying, "Come on in!" I now remember, I was around seven years old then. There was a game of marbles I used to love to play with other boys my age or so. At one time I counted over seven hundred marbles of various colors and sizes which I had won over a period of time! Storing these marbles got to be a royal pain in the butt! I began storing them in a cigar box but when that got full, I turned to empty jars until they too began to fill up. I finally ended up storing them in a small suitcase! When we were not playing marbles, we were doing other things.

One day we were burning trash, a normal task assigned to us from time to time. Well, one of the guys found a live round of ammunition. It measured about three inches long, so we figured it had to be a rifle bullet. Somebody suggested we throw the bullet into the fire! Like you know, there is always some smart ass in every group! So we did! Our curiosity was running wild to see what would happen next. We took cover just in case. It wasn't long before we heard a loud "POPPING" sound. We didn't know where it went but we thought that was pretty neat and cool and hot!

Another day we were walking down the street when we came upon a dead dog! Because of our curious and opportunistic nature, we just couldn't leave the poor dog there! To bury it would be too simple for us. Somebody came up with a better idea. We decided to take it to the shore. We got us a wheelbarrow because the dog was heavy, you know, dead weight! When we arrived at the shore, we all pitched in and at the count of three, we threw the dog in. We then waited and saw the tide take him in to the ocean. We really wanted to see the sharks devour him!

Going to the beach was an every-other-day event. It was really hard to stay away from it. I mean, not were the waters so calm, clear, and so sparkling beautiful, but there were Navy ships, sailboats, ferryboats, as well as yachts to watch while we ate grapes right out of the many grape trees in the sandy beach. The beach was only a short walking distance away. I recalled I didn't know how to swim yet, but that didn't stop me from going in and getting wet. Once in a while, the waves would rise and knock us down, defeating us like a high-priced boxer and pushing us to the dry sand. But sometimes we had other ways of fighting back.

My cousin, Abraham, was a fisherman. Sometimes he would invite me to go fishing with him. I loved his sailboat but hated going out to the ocean in it; as the saying goes, "Courage consists not in blindly overlooking danger, but in meeting it with the eyes open," quoted by Jean Paul Richter. The thought of the boat capsizing because of a sudden change of wind or a surprising storm always crossed my mind, and the fact that I didn't know how to swim yet! I recollect my grandfather's tales when he used to tell us that there is a fish out there in the ocean that save lives! I don't know about that! Although Grandpa never mentioned what type of fish it was that save lives, I figured it to be the dolphin. Yes, the dolphin was going to save my butt in case the boat ever capsized. But you know it did happen!

Relax, relax—I wasn't in the boat but Abraham was when a couple of years later he went out fishing as usual. The weather wasn't exactly nice that day, but Abraham, not respecting the ocean, like many of his fellow fishermen friends did, set out to go fishing that early morning. He said he didn't know how but it did happen! His boat capsized! His only alternative was to put a knife between his teeth and swim to shore. He swam for two hours! Well, guess what? No fish came to his rescue! I'm so glad I didn't go fishing with him on that day; otherwise I probably would have been "chopped meat" for the shark. I continued to enjoy my childhood years by playing baseball, hide-and-seek, and all those games kids play. Now that I remember, sometimes when we played hide and-seek, they never came to get me! I began to feel like my childhood joy was about to end.

According to Francis Bacon, "It is a miserable state of mind to have few things to desire, and many things to fear." I feared that my parents were getting ready to split. They were having problems getting along. I only desired for them to stay together and work things out, but it wasn't meant to be. After their divorce was final, Dad moved to the United States of America. He promised to send for me as soon as he could. My small little heart was broken because I didn't understood why I couldn't go with him! Out of four children my parents had, I was the oldest. I was seven years old. Following me were my two sisters and finally my brother. My younger sister was given to an elderly lady who was all alone and needed some companionship. She asked my mom if she would give her daughter to her.

My mom said yes and gave my two-year-old sister to her. This was a common practice then and up to this day I could never understand this practice. My other sister, brother, and I stayed with Mom, but not for long. Dad wrote a letter to Mom saying he was sending my uncle for me and gave her the date. Well, the minute I found out about it, I was "tickled pink" and was jumping up and down! I was so excited but was unaware that Mom was very saddened by my future departure from her. There was a calendar in the kitchen, and I remember every day I would cross out that day, thus counting the days when I would join my dad again.

Well, my departing date finally came. My uncle came for me. He said our trip would take us on a two-hour crossing the ocean by way of a ferryboat. Not only that, but another seven hours on an airplane! It was going to be the happiest day of my life! I was on my way to the United States of America, to a city my uncle said was called "Philadelphia."

CHAPTER II

PHILADELPHIA

My uncle said, "Philly is known as the City of Brotherly Love," the city where Dad went to live. I still couldn't believe I was finally on my way to meeting him and be with him. I had lost track of the last time I saw him. Now I could hardly wait for the Trans-Caribbean four-engine prop plane to land. The trip took almost eight hours! All I could remember about the plane ride was looking out the window of this flying machine, and because it was dark, I could not believe my very own eyes when I saw the city full of lights! I saw cars, so many rows of cars, so many headlights, following each other on the immense highway! This was Philadelphia. My dad had gone to live in a heavenly lit city and I already have fallen in love with it from the heavenly skies.

We finally landed at the Philadelphia International Airport. Upon stepping out of the airplane, I realized how cold it was. I had just left Puerto Rico, where the average temperature is 80 degrees, and landed here, where the stewards said it was 30 degrees! After picking up our luggage, my uncle signaled for a taxi to take us to his home. When we arrived there, his wife had set up a temporary bedroom for me to sleep that night. Although they claimed it was nice and warm inside, my bones were trembling from the cold outside. I must have remained cold throughout the night while I slept because in the morning my uncle's wife woke me up. She then asked me, "What the hell smells so bad in here?" It turned out to be crap! Yes, I had crapped all over myself!

After my initiation with winter was completed that night, it was time for the moment I had been waiting for a long time. The next day my dad came over to meet me. He said he couldn't come for me last night because he was working and it was impossible to get off work. He picked me up and took me into his arms and I hugged him hard! He needed a shave. I was so happy to see him. I told him, "I never, ever want to be left without you, ever again." He promised me that it would never happen again.

I stayed with my uncle a few months without having any more embarrassing nights. I was already getting used to the cold weather by now. My dad would visit me often and I was very happy. He had me so spoiled with lots of toys. I remember one particular Christmas Eve, my uncle sent me on an errand to the nearby grocery store. On my way back, I saw my dad going up the stairs with a red bicycle. For some reason, I didn't ask Dad about it.

The next morning, Christmas Day, I found the very same bike in the bathroom! Dad came over and said, "Santa Claus brought it to you!" I thanked HIM! I was in love with my new bike but it wasn't long before I found out that my dad was also in love! Yes, my father had fallen in love once again. Not only that, he was already making plans to remarry again. My happiness was rapidly fading away. I guess the bottom line was that I refused to share Dad with anyone else. I wanted his love to be just for me and me alone, but it wasn't going to be that way. He was to divide his love between me and her. It was going to be their second marriages for both of them. She had two children from the first marriage and of course my dad had his four, although not living with him at this time. After a short courtship, they were finally married. Afterwards, they settled down and purchased a small house on Wallace Street. It was a good beginning for Dad in comparison with when he first married my mom.

As I was told later, the downfall of Dad's first marriage was that at first he never provided Mom a place of her own to live in. Instead, it was always at some of Dad's relatives' houses. To complicate matters, my relatives on my father's side of the family never did accept my mom as being good enough for my dad. So they were constantly bickering and fighting over insignificant things. When Dad finally did get a place for Mom, the marriage was already on its way down. It was when he decided to go away to work in Saint Croix that Mom had an affair, which really ended the marriage. But this second time around, Dad was off to a much better start.

It wasn't until Dad had settled down with his new bride and stepchildren when I found out some startling piece of news. It seemed that on one of the conditions my dad's new wife agreed to marry him was only if he put his children in a foster home. However, this arrangement did not apply to her children. Dad was so in love with her, he naturally agreed to anything she said.

Well, it was only a matter of weeks before Dad brought me in to live with him and the rest of his new family. Dad was not done yet. My father went to Puerto Rico, and on a visit to see my two sisters, he brought them back to Philadelphia with him! After these last two additions to the family, all hell broke loose!

Our stepmother decided to take it out on me and my sisters for my dad having deceived her. She began to beat us up for anything that happened around the house, even though we were sometimes not at fault. Sometimes I thought I was being used as a punching bag by my stepmother. Many times my ears looked like cauliflower ears due to the result of these beatings. We would try to tell our dad our side of the story but our stepmother always said that we were misbehaving. My dad was always caught in the middle so nothing was ever done about the situation, ever. Because of such treatments, I ran away from home many times. Each time I did so, my dad knew where I went. However, each time he would take longer to come and pick me up. I usually ran away to my aunt's house, where I had a cousin my age.

We deeply felt that we were never welcomed in my father's house by our stepmother and her kids, just by our dad. But the saddest part of all was that our father did not believe us, or rather he chose not to believe us, when we told him that we were being beaten by his sweet wife. We had the bruises to show for it. Another thing, I never dared tell my dad that I was sexually molested by my stepmother's older brother. I was scared and feared that my dad would get in trouble with the law should he choose to take any action. Justice was served when I found out that this individual was murdered at his front home steps during a robbery attempt on him. I guess that was the reason I chose to finally move out and go live with my grandparents.

Living with my grandparents wasn't so bad after all. I went to school and although I was only ten years old, I had a part-time job in the evenings working in a laundry. The laundry I worked for was owned by a brother and-sister team. The sister would take care of the customers at the building where the actual laundry was located. Her brother, who was my boss, picked up dirty clothes from customers' homes and apartments. Our pickup days were Mondays and Tuesdays. Our delivery dates were Fridays and Saturdays. The last stop on Saturdays was a bar. I decided to take my shoeshine box and shine shoes for extra money! So while my boss spent his money boozing it up, I was making money on the drunks that tipped me well! I saved all the money I earned and I was glad I did.

My grandparents decided to move back to Puerto Rico. I definitely didn't want to go back to live with my dad as long as he continued to be married to my stepmother. So I then asked my dad's permission to go to Puerto Rico and continue to live with my grandparents. He said, "I don't have any money to send you with them." Well, of course I had the solution to that problem! I turned to Grandma and yelled to her, "Grandma, do I have enough money to go with you to Puerto Rico?" She answered back,

"You certainly do," with a big smile on her sweet face! I was given the OK to go to Puerto Rico with my grandparents. I was only twelve years old then. Deep sadness came to my heart because as much as I loved my dad, I was leaving him and who knew for how long until I would see him again.

My mom remarried and went to Philadelphia to rescue my other two sisters. She took them to New York City, where she now lives. Well, as for me, I was excited and looking forward to moving back to Puerto Rico again. I was ready for a change—or thought I was!

CHAPTER III

MONTE SANTO

Monte Santo (meaning Holy Mountain) is one of many small rural neighborhoods in the island of Vieques, Puerto Rico. I was twelve years old when I returned to the island with my grandparents. The best part about this trip was that my uncle was going to take care of me. He was my favorite uncle because he is an identical twin to my dad; it was like being with dad all over again! Of course, my uncle would not show me the same love that my dad would. Other than that, my uncle not only looks like my dad, he even laughs like him. The only difference between them, other than my uncle being a little heavier, is that my dad always worked for someone else while my uncle had always owned his own business. He has always made a good living from his business in order to provide food and shelter for his wife and their seven children, including one adopted little girl. He has always been a very generous man with a kind heart. That was why he agreed to take care of me, too.

My first month in Monte Santo was absolutely fabulous! I was having the time of my life but as always, it did not last for long. A month later school began and my nightmares began as well. First of all, I didn't know how to read or write in the Spanish language, although I arrived to school with excellent grades from fifth grade given to me by the Catholic school I was attending in Philadelphia. I was assigned to grade 6B instead of 6A. The smarter students were assigned to 6A. The lazier or slower learning students were assigned to 6B. My teacher made a deal with me. He said, "When you learn how to read and write your Spanish, I will transfer you to 6A."

I now remember how I hated when I was asked to stand up and read in Spanish. It was very difficult for me and at the same time it was very embarrassing. The other students would laugh at my mispronunciations of words. But on the other hand, when it came time for their turn to do their reading in English, I would return the favor by laughing out loud at their English mispronunciations.

Well, four months later I was worthy of being promoted to 6A. However, my teacher begged me to finish out my 6th grade in his class. He wanted to keep his insanity with the rest of the class. He said that as long as I proved to him through classwork and successful test grades, he would continue to teach to us, even if most of the class was doing terrible. In other words, his theory was that it was definitely not his teaching techniques that were bad but that the students were lazy or were not interested in school. I agreed to make his day, every day!

Being a new student in school had its advantages. I was treated as an American student. My clothes were different and to the students, my clothes were cool! It was because of this difference that a beautiful-looking twelve year-old blonde girl fell in love with me. However, we hardly had time to talk, so she decided for us to write each other letters. I had not yet even begun to learn to write in Spanish and was not ready for such written communications. So I turned to one of my newly acquired friends and he agreed to write the letters for me. Well, I don't know what he was writing but he did such a good job that the next thing I knew, he was holding her hand and walking her home! How do you like that! Some friend he turned out to be!

Girlfriend or no girlfriend, I had other friends who I look forward to spending time with each day, especially after school. Our best times were spent when we played baseball in the street. It wasn't in the main street so traffic was light. Most of our games were a lot of fun. We had girls play as well as boys. I remember this particular girl whose nickname was Blanky. Blanky was a good hitter but we always had to take cover when she came up to bat. You see, when she hit the ball, she would let go of the bat and it was anyone's guess where the bat was going to end up!

Incidentally, while I'm on this subject of the bat, which was made out of a flat piece of wood with a sharpened edge for a good grip, I remember this particular game we had. Our team began to argue with the other team over a disputed foul ball. My cousin Aida was on the opposing team. She called me a name I didn't like. So then I grabbed an empty can of tomato sauce and threw it to her. At this very same time, my uncle's adopted six-year-old daughter came over to see my cousin Aida. The next thing I saw was the can hitting the little girl's forehead and splitting it open. Blood began to come out of her forehead. My grandparents heard the cries of that little girl and upon investigating what happened, they took our baseball bat and gave me such a beating, I'll never forget! I since then have called that beating "La pela de mozo" ("The beating of a lad").

Afterwards I ran away from home, but where can you run to in a small island? I was only gone for about six hours and came back. Since I was now in school and things were starting to settle down a bit, I found out I was to pay a price for living with my uncle. You see, I was not to live there for free. Oh, no, I was assigned certain jobs in which I was to do in return for my food, room, and shelter. Along with my cousin, who was also twelve years old, we were shown these three-and-one-half acres of sugarcane field. It was now our responsibility for the maintenance of this field. I was to follow my cousin's instructions since he knew all about it from previous years of working on it. It seems that in order to grow good, healthy, tall sugarcane, usually eight to twelve feet and more, you have to till the soil with a hoe in order to clear the weeds. You have to plow early in the season and the soil has to be ridged to facilitate drainage. After the plant begins to grow, it has to be spread with fertilizers.

Finally, eight to ten months later, it is ready for cutting with a machete and loaded into the trucks for processing. Mind you, all of these procedures were to be done by my cousin and I with the exception of the cutting of the sugarcane, for which we had help.

One of my pleasant memories of Monte Santo was when my older cousin, Luis, invited me to accompany him to his job. Luis was some character. He would always find the easiest way out of his job duties and still get paid for it. He worked in one of the sugarcane fields owned by the government. Well, it was very hard work and he wasn't about to put in a full day working in the sugarcane field. Luis figured out an ingenious way of limiting his time in the field. He did it in such a way as to keep everybody happy including, of course, himself.

What Luis did was create his own job responsibilities in a manner that would definitely suit him.

My cousin convinced his Capata (foreman) to let him pick up, every morning, all the lunches of the workers, at their homes and brings them, nice and hot, to the sugarcane field! The Capata bought it and Luis became their lunch delivery man. All Luis had to do now was eat his lunch and work a few more hours in the field and be done for the day! Oh, and by the way, if you're wondering what method of transportation he used? I'll tell you. He built a square carriage with two wheels and hitched it to his horse! However, we had to contend with one minor problem. This horse had frequent gas attacks and they were smelly! While on the subject of this horse, my other cousin and I were in charge of getting grass from various pastures to feed this horse. One day we took a wheel cart and two picks with us to get some grass. We were so naive.

We decided to work on the same trunk of grass together by taking turns in hitting the trunk of grass with our picks, in a seesaw type of action. In other words, my cousin would swing at it once, then it would be my turn, then it would be his turn, then it would be my turn, etc., etc. We would do it in this manner in order to loosen the grass trunk quicker. Well, after a while I got mixed up in whose turn it was and I swung the pick. It wasn't my turn! The pick landed on my cousin's head! I broke his skull but he survived, but I wouldn't if he had not!

There were other jobs we were responsible for, such as feeding other animals, milking the cows, and so forth. However, the job that I hated the most, but on the other hand, loved the most, was fetching my cousin's horse. You see, Luis had a bad habit of letting his horse roam free but it was our job to go get him, which I hated. The horse was a beautiful golden stallion, and I loved to ride him once he was caught. He was wild in the sense that he was known to bite and kick, especially when he was among his mare friends. This was the time that I mostly and dreadfully hated: fetching him. The horse obviously wanted to be left alone with his friends and that was that!

One day my cousin and I set out to find the horse and finally found him near the lagoon, which was near the beach. This particular day my twelve year-old cousin said to me, "Go lasso him, he looks pretty tame today." I, the new innocent kid of the block, was a complete fool! I threw the lasso from a short distance away and landed on his neck. I then proceeded to do my normal, daily routine of walking very slowly towards him, and at the same time saying, "Staw, staw." Don't ask me what that meant! It must be horse talk, because everybody says the same thing to the horse to calm him down.

He looked pretty tame to me but like always, his ears would be standing tall and his big black eyes would be bulging with a precarious look on his face. I didn't like the look on his face either and was very cautious not to disturb him too much. When I saw that he seemed calm enough, I approached him and at the same time I noticed that my left hand was sticky from some kind of candy I was eating. I then proceeded to rub my left hand on his mane as a means of cleaning my hand. Then while attempting to mount him, without me seeing his face, he turned around and bit my neck! I'm alive today because he didn't take a good grip on my neck and I was able to duck away most of my neck. I remember running from this horse and climbing this impossible steep hill that led to the road from the lagoon.

I don't think there has ever been a human being that has been known to climb that steep hill without the use of ropes or other source of assistance, for that matter, but I surely did with my own bear hands. The horse was captured a few hours later by his owner, Luis, whom the horse feared tremendously. The horse was brought into the yard and was then tied to a 4x4 lumber pole. Luis then took a 2x4 piece of wood and beat him in the head continuously! My grandfather pleaded to Luis to stop beating the horse since the horse really didn't know why he was being punished due to the fact that the horse had bit me a few hours ago. Luis finally stopped the beating but it didn't make any difference to the horse.

A few months later the horse bit a little girl in the face and the same week he kicked my grandfather in the chest. My grandfather died a month later at the age of 86. The father of the little girl purchased the horse from Luis, and then took him to his barn and shot him in the head with a double-barrel shotgun.

Eleven months had gone by since I first arrived in Monte Santo. I was feeling pretty miserable, especially since the death of my grandfather. Certain things happened that didn't help matters. For instance, due to lack of beds, I was sharing my cousin's bed. However, he turned out to be gay and I refused to engage in homosexual activities. To each his own but it was not my cup of tea, as they say.

I wanted to leave Monte Santo but how? The opportunity presented itself a month later. It was during a New Year's Eve party. I came home and went to bed. I heard my cousin call me to her bed. There was no one in the house at the time. She tried to seduce me but I was totally inexperienced (a virgin) and therefore inadequate for the occasion. The next day she accused me of rape. Why me, I asked myself. All the things that were happening to me were beginning to get to me.

At this very same time my aunt was visiting us from New York City. She saw me and gave me some money. Later I found out that upon returning to New York City, she reported to my mother the condition in which she found me in. She told my mother that I was suffering from malnutrition, badly in need of clothes, and that my hair was full of lice. I myself didn't realize my condition as stated by my aunt. My mother then decided to come and rescue me and I was glad to see her! It didn't take much persuasion for me to decide to go back to live with my mother.

My uncle cried when I said goodbye. I insisted I didn't rape his daughter. However, to this day, her brothers still show animosity towards me. What did help my innocence was the fact that she cried "rape" again with an eighteen-year-old youth and they were forced to marry. This was the custom practiced then. I was just glad I was leaving with Mom to New York City.

CHAPTER IV

NEW YORK CITY

Shortly after my parents were divorced, Mom moved to New York City with my kid brother, who was my youngest sibling. She never let him out of her sight. He was the only part of our family that never got separated from Mom after the divorce.

Not too long after she arrived in New York City, she met a nice gentleman. She dated him for a short time and later accepted his proposal for marriage. It was his first marriage. He also didn't have any children of his own. Upon marrying and settling down, Mom was still not completely happy. She still had two daughters living in Philadelphia with my father and his wife. She constantly visited them and was determined to convince them to move back in with her. It didn't take too long for my sisters to accept Mom's invitation. Slowly but surely Mom was able to get all of her children back together again.

Mom; her new acquired husband, who looked like Rock Hudson; my two sisters; and my brother went to live in an apartment building, on the fourth floor.

The street: 116th Street and Pleasant Avenue. Let me just say that there was nothing pleasant about Pleasant Avenue. East Harlem, especially block 116th Street, was basically an Italian neighborhood. For obvious reasons, the Italians resented Puerto Ricans moving in to their neighborhood and especially block 116. However, an exception to all of this was my older sister. She was a very pretty-looking girl with slight blondish looking hair. She was friendly and was immediately accepted by the Italian young neighborhood kids on the block.

At this time, there had sprung up about twelve kids of various ages who hung around together all the time. I later found out that they were the local neighborhood gang. They called themselves the Red Wings. The Red Wings, upon seeing me for the first time, made me their enemy! I was thirteen years old, a loner, and paid dearly for it. They decided to call me "Chico"! I was constantly harassed by these punks. I could never get even because I was outnumbered 12 to 1. I always dreaded when my mom sent me on errands to the grocery store because I knew I would be meeting these guys in the hallway of the brownstone building where I lived.

One time on such an occasion, I was coming up the stairs with two bags of groceries. The smallest chubby kid harassed me and tried to pick a fight with me. I knew that one punch into his fat belly would leave him without air and send him to the wall but I resisted, thinking about the other possible eleven punches I was going to get from the rest of the gang! I always kept my cool but that only infuriated them even more. On this particular occasion they let me go through the hallway but when I got upstairs to my mom's apartment, which was four flights up the stairs, my jacket was on fire! I never saw it coming!

On another occasion, a Sunday morning, I was going to church with my younger sister. I was dressed for the occasion wearing a suit and tie. It was a very hot day in the summertime. Whenever such a day came, the fire hydrants would be opened up and the Red Wings would take over. That day they saw me and I saw them and I knew I was in trouble! They took me by my hands and feet and threw me into the fire hydrant, which was gushing water that reached across the street! I was given the drench of my life that day! My sister was spared.

Whether it was spring, summer, autumn, or fall, I would get it one way or another from these guys. This time it was winter and I was just coming out of school, which was across the street where I lived. It was snowing on that afternoon, I remember now. All of a sudden I saw a barrage of snowballs coming my way from the roof of the building above. I just couldn't understand why out of so many students coming out of school, I was the only one getting bombarded with snowballs! The Red Wings were at it again!

I survived all my teenage years living in Harlem. I finally did get to make friends throughout this time. However, I lost friends as well. The neighborhood was changing rapidly. Rivalry gangs or social activity clubs, as they preferred to be called at the time, were springing up all over the area. It was now the Red Wings on our block, the Untouchables on another block, the Black Knights further down the other block, all to contend with. We had a local cop whom we named Castro, who did his best to maintain the peace. He was well respected. However, it was getting pretty hairy around Pleasant Avenue.

A good friend of mine murdered another good friend of mine during a rumble. He was sent to prison at the age of eighteen! Other friends of mine were turning to drugs for a way to handle life. They were just living day to day. One day I had just bought a 16ounce soda pop and I met a friend, who had gone from a straight "A" student to a dropout junky. He asked me if I would give him a drink. I gave him my soda for a zip and he drank it all on one gulp!

Another time I met another friend of mine on the street. I was on my bike and he said he would buy my bike. He showed me some money. He said to let him ride the bike around the block and in this way he would offer me more money depending on how much the bike was worth. I gave him my bike to ride it around the block. You know what? That must have been the longest block I have ever seen because he never came back! I wonder if he meant to go around Manhattan but even there, he would have come back by now! I later found out he was into drugs and I never saw him or my bike again.

I also survived high school and finally graduated at the age of nineteen years old. Soon after, I landed my first real job working for an insurance company. I was thrilled about my first job until I found out that my brother was making more money than me and he didn't even finish high school. I was kind of mad at my school principal because he drilled to us students the idea that if we stayed in school and graduated, we would earn more money than those students who dropped out of high school. Was my brother an exception? I don't know about that!

Due to the little amount of money I was earning at this time, I decided to enroll in night college and take advantage of the insurance company's free tuition refund plan. Also although I was going steady with my fiancé at the time, there was no way I could afford to get married. A year later, at the age of twenty, I received a letter from The Armed Forces Induction Center. I was to report to White Hall Street, which was in downtown Manhattan, New York City. I was instructed to bring three days' supply of clothing. Enclosed in the letter was a subway token (my fare). Yes, I just have been drafted into the United States Army. I was thrilled and was looking forward to it, but was I in for a big surprise!

CHAPTER V

INDUCTION CENTER

November 18, 1965, a day I will never forget. On this day I was drafted into the United States Army. I just completed my 20th birthday. Attending Night College, having a steady job, thought my life was going great. Now Uncle Sam was taking over! From this day on, I belonged to the U.S. Army. I quickly found out that things are done the right way, the wrong way, and now there was the Army way, whether you agreed with it or not. As a matter of fact, even my thinking was taken care of by Uncle Sam. It all began that morning, November 18, 1965, at the induction center.

The center is located in an old building on White Hall Street in Lower Manhattan, in New York City. This was definitely the place not to be in. I reported promptly at eight in the morning with three days' supply of clothing as the induction letter stated. To my surprise there were a lot of guys there, my age or so, who were all strangers. I was hoping I would see somebody I would know, somebody I could talk to when permission was given to talk! The day began by all of us filling out these multiple forms. It was difficult for me to keep my mind on these forms and the information they were requesting of me. Instead, my mind wandered back to earlier that morning. It was very hard saying goodbye to Mom. She couldn't stop crying but I assured her I'd be back soon. It was the first time since I moved to live with my mom to New York City, eight years ago, that I was now going away from home.

My mind wandered back into the forms I was supposed to be filling out. I felt like a zombie! I forced myself to finish filling out these forms. Once we were done, we were then told to report to the medical department for a complete physical examination. I later found out that it was truly a complete physical exam. We were all given a full set of vaccinations. Also our blood samples were taken and submitted to the laboratory. Our medical records were filled out to their entirety, from allergies to Zymosis diseases.

We were then told to line up in a straight line and strip naked from the waist down. Well, this was very embarrassing to me. I have always been a very private person. I'm sure I wasn't the only one feeling this way. The next thing I heard was this loudmouth sergeant hollering from the top of his lungs, "On the double!" We all jumped up and stripped in a second! If this wasn't embarrassing enough, we were then told to bend down! This time we did it on the first command by Old Loudmouth Sarge. The next command really got to me. "Spread out your cheeks!" By this time I was so nervous that I began to develop a bad case of stomach gases. I then took a peek at Old Sarge and to my astonishing surprise, he was coming down the aisle with a medical doctor. I couldn't believe my own eyes what I saw next. The Old Sarge and the doctor were checking out our rear ends with a flashlight! I just couldn't figure out what they were looking for. Whatever it was, I didn't have it.

By the time they got to my rear end, I just couldn't hold it anymore and I let one out that sounded like loud thunder! The sarge turned towards me, bending down to my level, and said, "What's wrong, son?" I said, "Nothing, Sergeant, I'm just nervous." He then said, "Try to contain yourself for our sake." By this time all the guys in the aisle were turning red from holding back from laughing until they couldn't hold back anymore and finally burst out laughing. By now I wished there was a hole in the floor to disappear to. The sergeant was dead serious by now and his face was turning red with anger and frustration. He mentioned something to me about an Article 15. Of course I didn't know what he was talking about.

He then said, "I'll pretend this never happened and will forget it this time." I said, "Thanks, sir." He then yelled back at me, "Don't call me sir, I work for a living!" Again, I must say I didn't know what he meant by that but I decided to keep my mouth shut. I was already in too much trouble as it was. The last part of our physical exam consisted of questions asked to us. For instance, we were asked: Is there anything wrong with us that we feel would hold us back from performing military duties? I hated to raise my hand, since no one else did, but yes, I did raise my hand. The sergeant looked annoyed at me and said, "What is it now?" Since I always thought I suffered from a heart murmur, I said to him, "I have a pain in my heart." He said, "Oh, you do. Come up front and jump up and down ten times."

I have to admit I felt pretty stupid but I did as the sarge said. He then felt my heart beating very fast and said, "You'll survive, now get back in line!" By now it was time for our first military "chow." To say the least, I don't remember what was served to us that day. I do remember making an "UGGGHHH" sound! We were very hungry and didn't have much choice but to at least attempt to eat some of it. Not that it mattered much because my stomach felt it had butterflies in it, not knowing what to expect next on this day. After chow we were to report to a large room, where we were to be seated and wait for further instructions. It was another of this "hurry up and wait" situations that I was already getting accustomed to.

Finally a big, rugged-looking sergeant came to the room with a clipboard. He said, "Listen up, men. The following fifteen names have been drafted into the Marine Corps." My heart was pounding so loud I thought it was going to burst! He began reading last names first, beginning from the letter A until Z. The letter T was skipped and I knew I had survived that ordeal. We were then told that orders were being cut for us for our next place of destination. Let me just say at this time that as far as military terminology, ranks of command, and so forth, I was totally naive. I never joined The Boy Scouts of America when I was young or had joined any organization like ROTC, which exposed me to any of these military procedures.

As far as identifying ranks, I only knew that a general was identified by a star on his shoulders and a sergeant by his stripes on his side of his uniform. This was all new to me. So when the sergeant said that orders were being cut for us, I didn't know what he was talking about. All I could think of was scissors to cut orders—whatever that was! So I turned to the guy next to me and asked him to explain to me about these orders the sarge was talking about. He then explained everything to me. From then on that guy next to me became my best friend. His name is Robin Brown.

By now we were told to report to the mess hall at 18:00 hours!? I found out from my best friend, Brown, that the mess hall was what the cafeteria was to civilians. The 18:00 hours was military time, translated to civilians was 6:00 P.M.! After chow (again, UGGHHH), we were told to report to the same room we were before chow, to receive our orders. Our freshly cut orders were given to us. I read mine and I thought they were written in another language! They were typed and all I read was a bunch of numbers and short phrases.

However, I did notice that the orders stated that I was to report to Fort Jackson, South Carolina, by way of the rail train. I asked Brown about his orders and he said he was also assigned to Fort Jackson,

South Carolina. According to my orders, the U.S. Army was transporting me to South Carolina by way of the rail train! The closest I ever came to riding on a train was the New York City subway. I was looking forward to it. What I didn't know was the experience I was to have on that train!

CHAPTER VI

TWENTY-HOUR TRAIN RIDE

During my teenage years I made use of the New York City subway system like millions of other people in the city. A schedule of trains and times were met but as always, there were sometimes delays due to traffic congestions, breakdowns, and so many unexpected problems that are bound to happen on a day-to-day situation. The most time I have ever spent on a subway from one destination to another was an hour and a half! I'm pretty sure subway commuters have taken a little bit more or less but time was of the essence!

No matter what time I spent riding the subway, I never minded much. There is always something happening while on your ride to your destination. As an example, I remember a blind old man walking through the center of the aisle with his walking stick as a guide. He guided his stick with his right hand. In his left hand was a tin cup in which he collected donated money, mostly coins. While bouncing his tin cup, a quarter fell to the floor and he bent down, picked it up, put it back to the cup, and continued walking! Time spent on a subway or passenger trains are very much similar, or so I thought. I was in for a big surprise!

I received my orders to report to Fort Jackson, South Carolina, by way of railroad train! I had never ridden in a passenger train. I compared it to be simply an extended subway ride but was I wrong! We boarded the train at exactly 19:00 hours! It was a passenger train that must have been chartered by the U.S. Army because I didn't see any civilians on board. The passengers were all young men like me.

With us on the train were a few young soldiers, smartly dressed in starched khaki uniforms. On top of their heads were green hard hats with the letters MP inscribed in bold letters. They had a decorative white rope around their shoulders. On their necks they wore maroon scarves that covered all of their necks. On their right side of their belts were holsters with a gun! They had boots with the glassiest shine on the tips that I had ever seen before in my life! These guys were very impressive-looking soldiers but at the same time, they appeared, to me anyway, that they were tough, like not to mess with them, if you know what I mean! I just wished I looked as sharp as they did!

Later on I found out who these sharp-looking soldiers were. Like I said before, I was totally naive as far as the military was concern. I found out that these soldiers were wearing helmets and not hard hats! The letters MP inscribed in their helmets showed everyone, except me, what they were! Yes, they were the Military Police. They were with us to keep the peace on the train en route to South Carolina. I began to feel like a prisoner sentenced to two years, serving Uncle Sam, stuck on a train! Although we had boarded the train at 19:00 hours, it did not begin to move until quite sometime later. Later I found out that in military time: "It's hurry up and wait!"

While we were waiting to get going (eventually), we were assigned our rooms or quarters as I later found out was the military term properly used. Again I felt like I was in jail! I only know of jail because of watching movies with convicted felons in those jails. I could not believe how small these quarters were! They were very small with double bunkbeds for sleeping. These bunkbeds would sly into the wall when not in use. My sleeping quarters was so small, I began to be depressed. I couldn't stop thinking how big my bed was at home sweet home!

The train finally began to move! (Hurrah)! At first it began to move very but very s l o w l y! Then after a few sudden jerks back and forth and a few starts and stops, it finally picked up the pace but not for

47

long. It began to slow down again and again! During the entire train ride, I kept waiting for this train to pick up speed and maintain that speed but it never happened! Where were the fast trains that we all saw around us?

I began to get awfully bored and I knew I was not the only one! To stay in that confined room for this long amount of time would drive anyone crazy along with a bad state of depression! So I did what everybody else did. I stepped out of my (cell) room and began to mingle among the would-be future solders. I found myself exploring the rest of the train. I found among the various cars of this slow train was the lounge car. In the lounge car I found food to be eaten (for free!). This was where I found most of the crowd to be hanging out. Some were playing cards while others were just hanging around. The bottom line was that we did anything to keep us from thinking about the long train ride we were experiencing. It was a very slow ride. Even when we looked out the window, there was nothing there to see except desert and desert lands.

Among the many stops and go we made, we did make one very long stop to deliver mail to Washington, D.C. We then proceeded to our final destination: South Carolina!

CHAPTER VII

SOUTH CAROLINA

November 21, 1965, I arrived at Fort Jackson, South Carolina. It was a very cold day and nasty weather, but that did not stop the men in uniform marching down the streets in complete unity. Every soldier's feet and hands moved in complete unity. Their faces and eyes looking straight ahead. I'd never seen anything like it. Boy, oh, boy, they were looking sharp as could be. In the front of the marching group was a soldier carrying a banner indicating the name of company they represented.

Something else intrigued me. One soldier on the side line, marching by himself, would sing a verse of some kind and the rest of the marching soldiers would respond back. From time to time, the lead soldier on the side line would sing: "Sound off, sound off," and the rest of the marching soldiers would respond: "One, two, three, four." It was so impressive to see these soldiers marching and singing in such a disciplinary manner! I just wished they were on television so that the viewers would witness how great they look and sound.

I myself began to feel very proud that I too was part of these soldiers. I did not have the training to march like them yet but in due time I would be looking forward to doing so. As a matter of fact, I found out that once you are assigned to a platoon, all soldiers in that platoon do everything together. They march to the barracks, to chow, to P.T., and so forth and so on. As I later found out, Fort Jackson was my processing base. I was not going to be assigned here permanently.

We were assigned our temporary barracks for the time we were to stay in Fort Jackson. I can only say of my first night in our assigned barracks: Oh, what a night! Here we were, a bunch of young men, each with their own beds, separated by just a few feet apart from each other. We had a representation of many ethnic backgrounds of races. There were men of white color, black color, brown color, yellow color, you name it, and we had it! There were Puerto Ricans, African-Americans, Irish, Italians, and Asians like Chinese, Japanese, Koreans, and Hawaiians. They were all there. We began to speak to one another, sort of like to introduce ourselves. I had already introduced myself to an inductee, like myself, to someone I met at the induction center, in Manhattan, New York City. On my induction date, November 18, 1965, I introduced myself to Robin Brown.

Brown was African-American, 6'2" tall, weighed around 200 pounds, and fit the job of a bouncer in a club or anywhere you'd see him. He was from Manhattan, New York City, like me. He was also of Christian faith, just like me. While I was talking to Brown, a soldier across from us hollered and yelled to me, "Why are you always hanging around that n——-?" I was shocked when he used the N word. I yelled back at him and said, "Brown is my best friend!" I later found out he was known as a redneck. I didn't know about that. His neck was as white as snow! As soon as the redneck fellow shouted the N word, all hell broke loose! All the African-Americans in the barrack began shouting, "Rebels!" The Rebels shouted back, "Yankees!" I didn't know what was going on. I thought the Civil War had ended a long time ago! I told my best friend, Brown, "Don't mind them, we Yankees stick together no matter what!" As the Rolling Stones song says, "We got to get out of this place!"

I lasted eight days in Fort Jackson, South Carolina. On my last day, I remember falling in formation early in the morning as usual. By now it was all routine and I was getting used to it. However, this morning was different. We were all at attention. I then heard the D.I. (Drill Instructor) yell, "TORRESSSS, MICHAEL AAAAAAA." His accent was so southern that I could not understood him. He repeated my name one more time. I began to wonder who he was calling. Now the third time he called the name, he called it in sort of like in slow motion: "TORREESS, MMIICCHHAAEELL, AAAAA!" By now I realized he was calling me! I raised my hand out of about 100 soldiers in formation. He yelled, "Front and center, now!"

I reported front and center like he said. He then approached me and his nose was about an inch from mine! I was about to crap on myself! He then shouted at me, "Soldier, did you take stupid pills this morning?" I responded, "No, Sergeant!" He yelled at me again, "I don't hear you!" So this time I shouted back, "No, Sergeant!" He then yelled at me again, "Report to the company headquarters and see the office clerk." Before reporting to the company clerk, I headed straight to the bathroom! I reported to the company headquarters clerk on duty. He handed me my new orders. I was to report to Fort Sam Houston, Texas, for my basic training.

This time I was not traveling by train. Along with my orders, I was given a flight ticket to Dallas, Texas, with connecting flight to San Antonio, Texas. I was excited!

CHAPTER VIII

SAN ANTONIO, TEXAS

November 29, 1965, I arrived at San Antonio, Texas. Texas is known as the Lone Star State. It took us a couple of hours to fly from South Carolina to Dallas, Texas, and then another hour on our connecting flight to our final destination, San Antonio, Texas. On my flight to San Antonio, Texas, I met my best friend, Robin Brown. He too was being assigned to Fort Sam Houston, San Antonio, Texas for basic training. This was my first time in the state of Texas. Actually, come to think of it, my first time anywhere other than Puerto Rico, Philadelphia, New York City, and Niagara Falls, New York side.

I began to notice right away that in the military, you get to be assigned to many places and very frequently! For someone like me, I was getting to enjoy it a lot! For now I was enjoying the beautiful city called San Antonio, Texas. One of the things I noticed right away was these streams of water, like creeks of water running through the middle of town. It was breathtaking! They were very nice to see and at the same time a sense of tranquility! Another thing I found out is that San Antonio has a large population of Mexicans. Now it just might be part of their culture but they love their trucks! I noticed that while they were driving their trucks, their girlfriends or wives were sitting very near to them.

You always had room for a third passenger on the front seat. Many times I observed these couples sitting very close to each other while driving on their trucks. I found this to be awesome, unique, and at the same time, to be very romantic! A place that was definitely not so romantic but was full of history was the Alamo. I visited the Alamo because I've read so much of its history. The walls are still standing there but its history continues on for generations to come and go. I'm so glad I was assigned to San Antonio, Texas, for my basic training. As my orders stated, I was to take my basic training in Fort Sam Houston, San Antonio, Texas.

The weather in Texas, in November, is so different from Philadelphia or New York City. Here in the mornings it gets bitter cold. In the afternoon it gets warm, like in the summertime. At nighttime, it gets very cold again.

My basic training began with our assigned sergeant coming into our barracks at 05:00 hours to wake us up. Let me describe this sergeant. He look and spoke like Dorky Buckteeth Mater, the greater in the cartoon character movie. He had a set of buckteeth just like Bugs Bunny! He would wake us up by hitting a large tin pale with a stick, making a lot of noise. I called it making a lot of racket! I thought immediately he was deaf and looked very stupid! He gave us fifteen minutes to make our bunks (bed), put our fatigues on, brush our teeth, and fall outside in formation. He wanted us outside, on the double, to fall in a single-line formation.

While in formation, I heard this Mexican sergeant with a very bad Spanish accent, yelling at us, saying, "Are you men in attention or areeze (at ease)?" I only understood part of what he said. I didn't want to face another sergeant that I couldn't understand what he was talking about. I still had a fresh memory of the sergeant with the southern drawl accent in South Carolina, sticking his nose very close to mine, asking me did I take stupid pills this morning! Now I was stuck with this Mexican sergeant's Spanish accent! I was already getting prepared for this Mexican sergeant. Should he call me out of formation to report front and center, I was planning to keep my nose as far as possible from him. If he attempted to talk to me and I couldn't understand his English, I would request for him to speak to me in Spanish. I just hoped his Spanish was better than his English.

Basic training became routine to us. We were woken up at 05:00 hours every morning. P.T. (Physical Training) began at 06:00 hours. P.T. consisted of a lot of repetitions of pushups, sit-ups, jumping jacks, all kinds of stretches, you name it, and we did it. We also did some running. Sometimes we ran for a mile or more. Other times we just jogged. After all these various exercises, we were ready for chow. Now chow consisted of potatoes three times a day. It was served to us in various ways. For breakfast we might have eggs scrambled, hardboiled, or sunny side up. Along with eggs, we would get toasted bread and strips of bacon. Potatoes for breakfast would be diced potatoes. We had juice and coffee. Sometimes they threw in a fruit, like an apple or a banana. Other times they might offer us oatmeal or cereals or grits.

As for lunch, we might be having hamburgers, cheeseburgers, French fries, milk, soft drinks, or juice. For dessert, we had a choice of apple pie, an apple, banana, strawberries, or a pineapple. Now for dinner, it was steaks or pork chops, with mashed potatoes. Included in the menu were sweet carrots, spinach, beets, sweet peas, a salad of lettuce and tomatoes. We were given a choice of soda, milk, or juice. The chow was not bad at all. The main objective of basic training was to convert ordinary men into disciplinary soldiers. This conversion included physically as well as mentally.

One of the first things the military did to all of us men was to shave our hair off our heads. We then looked all alike as far as our heads were concerned.

We were issued our clothes, namely, our pants, shirts with our last names printed on them, fatigue jackets, ponchos, and boots. We were given adhesive labels with our names to be installed inside our boots for identification purposes. Every soldier was issued a silver chain along with a silver ID tag. On the ID tag, it had our names, service numbers, blood types, and religions. I later found out that these silver tags were called dog tags. Now I felt like we were being treated like DOGS!

On my dog tags as for my religion, it stated "Pentecostal." In other words, I'm Christian. Now when I was drafted into the United States Army, I was asked in my paperwork forms, what religion, if any, I belonged to. I wrote down "Christian-Pentecostal." At this same time, while filling out my military forms, I declared myself as a C.O. because of my religious beliefs. C.O. stands for Conscientious Objector. What is a Conscientious Objector? A C.O. is a soldier who will gladly serve his time of active duty in the U.S. Army with some conditions. Due to my religious beliefs and Christian faith, I refuse to harm another human being in any shape or form.

For this reason, I refused to accept a firearm to be used to cause wounds or even kill another human being. Added to harming another human being, I refused to engage in hand-to-hand combat as well. I understood that I was now a soldier of the U.S. Army. However, I was drafted into the U.S. Army. I did not volunteer into the U.S. Army. I had the right to serve my country as a C.O. soldier.

Part of basic training included familiarization with firearms such as a .45 pistol and perhaps an M14 or M16 assault rifle. As I indicated before, I refused this training and because of it I went through hell! I was ridiculed; I was called a coward, a chicken.

When soldiers passed by me, they made chicken sounds. I was even called a traitor to my country! I also refused hand-to-hand combat training. The D.I. was furious with me. He said, "Torres, you know what is going to happen to you? The Army is going to send you to Medical Training and you are going to become a medic, and that's not all! The Army is going to send your ass to Vietnam!" He tried to convince me to take the training and to learn as much as I could on how to defend myself. He then said, "In Vietnam, Torres, the bullets are for real!" I responded back to him, "Well, my Lord is for real too!" He then shook his head left and right and, pushing his hand away from me, went away.

I continued my basic training as usual. One aspect of training consisted of completing an obstacle course. The obstacle course consisted of various challenging tasks to complete. In one task we had to crawl through low barbed wires while we were under simulated cross enemy fire! Another obstacle task was to cross a water running creek using a single rope, using our feet and hands upside down! Among us was a Hawaiian soldier who weighed around 200+ pounds and was 5 feet 4 inches tall. He began to cross the creek upside down, as instructed. Well, halfway across, he somehow slipped his hands and was left dangling. He tried to lift his heavy body up to grab the rope but was unable to do so.

After a minute or so went by, he gave up! He then stared at the D.I. in charged and yelled, "GERONIMO," closed his eyes and opened his legs, and fell hard to the creek below, which was full of rocks! He was very badly bruised up from hitting those rocks. He survived the fall! The medics took care of him. I will never forget that fall! It should have been recorded and sent to *America's Funniest Home Movies*. It would probably have won the top prize money! Besides our physical training, we also had classes to attend to. One of our classes was all about marching and its commands. We were instructed how to snap into attention, the proper way to snap to at ease, the correct way to do a left face and a right face.

All commands were to be done in a snapping type of way all the time. The instructor was Mexican but he spoke without a Spanish accent, which was great for me. He performed for us a perfect about face. He expected us to do the same when our platoon leader gave us the command of about face: HUH! I didn't know why they hollered "Huh" after each command but we got used to it. Once in a while our Mexican instructor would get off the topic and would talk to us about a war story when he did his tour in the Korean War. I found it to be fascinating. Maybe perhaps someday I will have my own stories to tell someone who is willing to listen!

"Listen here, men," the sarge would yell to us during our final formation. You see, we were finally graduating from basic training. HURRAY, we survived all the humiliation, verbal, physical, and all the harassment, thrown to us. We took it like true soldiers we now fully deserved to be called! Graduation was great. All the soldiers were so happy. I was happy, too! My best friend, Brown, was happy! We said our goodbyes to our P.T. sarge and they wished us all "GOOD LUCK!" Now, I asked, "Why good luck?"

Well, a while back you might recall a P.T. sergeant predicted to me that because I declared myself to become a C.O. (Conscientious Objector), I would probably be sent to medical training to become a medic. Well, I found out that he was absolutely right! I received my orders for my A.I.T. (Advance Individual Training), to be done and located at Fort Sam Houston, San Antonio, Texas. So I received good news and bad news. The good news was that my AIT would be here in Fort Sam Houston, Texas, where I just completed my basic training. The bad news, to me anyway, was that I would be training to be a medical corpsman.

Basic training was all about getting our bodies into the best physical shape we could be. As far as mentally was concerned, the D.I.s made a point of stripping us of our personality. Their goals were to brainwash us into thinking, talking, looking, marching, and acting like a soldier. Now A.I.T. is so much different from basic training. A.I.T. was all about attending classes. It was all about medical training. There was so much training and so little time to absorb everything our instructor was teaching us that many times we felt overwhelmed. From time to time the instructor would drop a hint that we would need to apply all this training should we be assigned to Vietnam!

The training was so intense that the stress was getting to me. To release my stress, I decided to do something about it. Now with A.I.T., we were given time for ourselves. We spent our day attending classes. However, we were free in the evenings. So a flyer was passed around the barracks, asking soldiers to sign up for the softball league that was just being formed. Back in the day, I remember playing baseball in the local neighborhood parks. I also played for my high school. I'm a natural switch hitter and because of my strong arm and speed, I did very well in the outfields. Mind you, I'm only 5'6" tall but I got around quickly, whether outrunning line drives or running the bases.

I always had this passion for playing ball. So like I said before, to relieve stress from all the medical training, I decided to join the softball league. The coach asked me what position I would like to play. I responded, "Left field or right field would be fine with me." So the coach assigned me to right field, where I played in the beginning of the season. I remember this particular defensive play I did. The hitter on the opposing team hit a line drive to right field. He ran to first base and continued running to second base.

It was a typical double for the books but to my surprise while I was picking up the ball, the runner continued running to third base without stopping at second base. I in turn shifted my body into my throwing position and fired a strike to third base, the throw beat the runner! He never had a chance to reach third base! He was tagged out for the third out of the inning. When I arrived at our dugout, all of my teammates were giving me the high-fives. It felt pretty good to contribute defensively to the game. By the way, we won the game!

We continued to play well as a team that we created a reputation for ourselves! Once the opponent team found out they were playing against "The Medics," they knew they had their work cut out for them. The league included soldiers from Lack land Air Force base, which was on the other side of San Antonio, Texas. As I said before, our team's name was "The Medics." We had beaten every team we faced thus so far! Around the middle of the season, our catcher came down with a bad case of appendicitis. His appendix had to be removed. Afterwards, because of his required recuperation time needed, he quit the team.

Our coach had extra outfielders but no catchers. I asked the coach to give me a chance at catching. He already knew I had the arm as well as the speed! We had two good pitchers who I had to work with. Even though I'd never been behind the plate as a catcher, I caught on very quickly. As a base runner on second base, I always observed the opponent catcher giving signals to his pitcher. Now that I was a catcher, if I was on second base as a runner, I tried to predict the pitch being thrown to my hitter, depending on the catcher's signals to his pitcher. I would relate through signals to him the predicted pitch coming down! Sometimes I was successful and sometimes not.

I fell right into the position of a catcher and I loved it, especially when the opponent team attempted to steal bases on me. They soon found out about my strong throwing arm. In softball, there is no leading as a runner on the bases. So picking off a runner on base does not exist like in baseball. The runner is allowed to run only after the pitcher releases the ball. We were lucky in the sense that another player joined our team and he was also a catcher. So it worked out, being he replaced our injured catcher that quit because of his appendicitis. So for the rest of the season we shared the position of catching. However, out of the two pitchers we had, one of them specifically asked the coach for me to do the catching. I was honored to do so. Maybe it was because I learned quickly how to frame pitches.

In other words, when a pitch was close enough to be a strike or a ball, I always tried to catch the ball by moving my mitt towards the plate all the time. Chances were the umpire would give me the strike call. This was why hitters argued with the umpires so much, and they had every right to do so, because to them that particular pitch WAS a ball!

The season was a success to a point. We were 18-0. Now why do I say to a point? Well, we reached the championship game. I say the best pitcher for us was pitching and he requested me to do his catching. Our opponent's pitcher was their best. On both sides was innings of strikeouts, ground balls, and fly balls.

However, the opposing team hit back-to-back hits by their 5th and 6th batters on the top of the 5th inning, scoring one run by the time three outs were made. So they scored first and led 1 to 0. Their starting pitcher was still pitching. The pitcher then pitched to our 4th batter in the 5th inning and he grounded out to second base for the first out. Then our 5th, 6th, and 7th batters all hit safely to load the bases. Unfortunately our 8th batter hit into a double play and we couldn't score. He hit the ball hard to third base. The third baseman stepped on third for the second out and then threw home for the third out.

There were no runs scored by either team by the 6th inning.

We were up batting on the top of the 7th inning. Our pitcher, hitting 9th, struck out for the first out.

Their pitcher then made a big mistake and pitched a slightly high pitch, in the letters, as they say. I know he didn't want that pitch to land where it did. So our leadoff batter jumped on it and took it downtown for a HOME RUN. We were now tied 1-1. Our next batter flied out to third base for the second out. Our next batter flied out to right field for our 3rd out of the top of the 7th inning. Our opponent's bottom of the 7th inning began with a pinch hitter batting in the spot of their pitcher, their 9th batter of their lineup. I gave my pitcher the sign for a curveball inside the plate. He responded with a curveball as I requested. The umpire called it a strike even though the batter opted not to swing at the pitch.

As for the next pitch, I requested another curveball but my pitcher turned his head, giving me a no. I requested a fastball and he said yes by nodding his head up and down. He pitched his fastball and I could see he put something extra on his velocity because it came down extra fast. The hitter decided to swing at it. He hit the ball, but the way to describe it, we called it a swinging bunt. The ball took off his bat towards our third baseman that was playing a little deep. The ball took off in a speed of in the middle between a bunt and a ground ball. I couldn't get to it. Our third baseman also couldn't get to it. So it was up to our pitcher to make the play. The runner had good speed. Our pitcher went for the ball, grabbed it barehanded, turned towards first base, and threw the ball a little high. It was playable by our first baseman to catch it. The runner beat out the throw by a step.

The top of the order came up to bat. With one runner on first, I expected a sacrifice bunt. I called timeout to talk to my pitcher. We both agreed that a possible bunt was coming down. Our manager joined us at the mound, along with our third baseman and shortstop. The manager wanted the infield in for this bunt play. I requested my pitcher to pitch a low pitch for a ball count. It would be a pitch that only a low ball hitter would swing at. I went back to home plate. My pitcher threw the low pitch as I requested. The batter squared off to bunt as we all expected. He did not swing at the low pitch. The umpire called the low pitch a ball for the count. I then noticed that the batter, while getting ready for the next pitch, was crowding my plate. I requested my pitcher to pitch, as we called it, a brushoff pitch. I knew it was going to be called ball two by the umpire.

My pitcher pitched a beauty of a brushoff pitch. The hitter hit the ground hard while avoiding getting hit by the ball. The umpire called it ball two. For the third pitch, I requested my pitcher to throw me an inside curveball over the left side of the plate. The hitter was a right handed hitter. My pitcher brushed off my request. I then requested a fastball and he agreed by giving me the yes sign. My pitcher threw a letter-high fastball. The runner on first base took off as soon as my pitcher released his fastball. The hitter bunted the ball between my pitcher and first baseman. I did not have a shot at the play. The second baseman reached the ball first and his only play was to throw to first base for the out. Our shortstop covered second base. The runner slid safely into second base.

Up to bat was their second hitter on their lineup. He was a switch hitter like me. He chose to bat lefty. He chose to hit lefty to make sure to advance the runner on second base by putting the ball in play at the right side of the field. We decided to pitch to him inside, outside, hoping he would swing at pitches hittable to our favor. We needed a ground ball to the infield for a possible double play. That would be sweet. The hitter went to a full count of three balls and two strikes. The hitter hit a fastball to deep right field, in which our right fielder made a spectacular catch by leaping way up on the fence and grabbing it for the second out. When he made the out, he fell to the ground. The runner on second base tagged up and ran to third. I'm pretty sure the third-base coach saw our right fielder fall to the ground and decided to gamble and send the runner to tag up and run to third base.

Our right fielder, obviously knowing the situation, quickly got up and fired the ball to third base. It was a good throw, bouncing once, before reaching our third baseman. The runner slid away from the third baseman and grabbed the bag with his left hand, thus completing his tag-up successfully. He was called safe by the umpire. Now with two outs and a runner on third base and in scoring position, we asked for a timeout to the home umpire. Our manager brought the infield to the mound for a meeting. The manager went over our strategy for the coming hitter. Our pitcher was going to pitch only low pitches, along with low curveballs. The infield was going to play in tight. Their hitter, third in their lineup, came to the plate.

He was a right handed hitter.

Like the previous hitter before him, we did not want to pitch anything hittable to him. If he hit a pitch, hopefully it would be a grounder for the third out. Well, the hitter turned out to have a good eye for the ball. We tried to see if he would swing at some low pitches, low curveballs. We then tried to pitch to him some high pitches. He ran the count to three balls and two strikes for a full count. I asked for a timeout to the home umpire. I wanted to talk to my pitcher. I asked my pitcher, "How do you feel?" He said, "I'm fine!" I asked him what pitch he felt like pitching next. He said, "Low fastball down the middle for strike three." I said, "Go for it!" I went back to home plate to play ball! My pitcher then threw a fast lowball pitch that to me was unhittable.

The hitter, however, thought otherwise and thought it was hittable. He hit a hard ground ball to our second baseman. The runner, who was on third base, being that there were two outs, took off towards home. The ball must have hit a pebble or something on the ground because the ball took a high hop and hit our second baseman on the chest. The ball fell to the ground. He had seen the runner taking off towards home so he picked up the ball and I got ready for a play at home. I blocked home plate and waited for his throw. He hurried up and threw me the ball a little high, not what I was anticipating. I was able to catch the ball. When I tagged the runner with my mitt and ball, the runner was sliding home. He extended both of his hands and knocked the ball out of my mitt and the ball fell to the ground! He was called "SAFE" by the home plate umpire.

I argued the call as well as our manager. I told the umpire what the runner just did. Apparently the umpire did not see what did happen! We lost the championship game, by the score 2 to 1. Our record for the season was 18 wins and 1 loss. Our base commander was so proud of his MEDICS having such a successful softball season. Playing softball was a lot easier for me than medical training. We continued our medical training. We learned how to administer injections the proper way. I will elaborate more on this topic at a later time. The instructors brought us dummies with various fake wounds over their bodies so that we could get used to see them and practice on them using proper procedures to treat them. I will also go into this topic in more details at a later time. You will never believe it!

The time finally arrived and we finished our Advance Individual Training (AIT) in the medical field and we had earned our title of medics, also known as Medical Corpsmen. While attending our graduation, emotions were very mixed. You see, along with graduation certificates, we were also given orders to where to report for our next assignment. Some medics were assigned to stateside duties. As I mentioned before, a good buddy of mine received orders to report for duty to Anchorage, Alaska! Well, you guessed it. We just made fun of his assignment. We all agreed that he was going to freeze his butt! However, had we hindsight where the rest of us were being assigned, we would envy his assignment.

Other graduate soldiers elected to go air born, where their continuing training would be at Fort Bragg, North Carolina. I would meet some of these soldiers again but under unfavorable conditions. Because the demands for medics were so high, the majority of graduated medics were assigned to South Vietnam, including myself! I and everyone I knew on that graduation group were devastated. We were all ordered, in this group, to report to South Vietnam for a one-year tour of duty. We were given 21 days of leave time before reporting to our overseas duty. Time was going to be short and would go fast. I took a flight to New York but not for long.

When I arrived at Kennedy International Airport, New York, I went to visit my mother, who lived in Manhattan, New York City. Upon visiting my mother, I asked her about a girl I was very much interested in but never asked her out on a date. I remember going to her 9th grade graduation way back then, and that was that. My mom said, "Do you mean Luz Elba?" I said, very excited, "Yes!" Well, she said, "Luz Elba moved back to Puerto Rico, where she is now living with her parents." I was shocked! I then said, "Mom, I have only three weeks before I report to Vietnam. I have to see her before I go to Vietnam." I didn't even know in what area of Puerto Rico she was!

I was immediately able to contact by phone Luz Elba's older sister, who lived in the Bronx, New York. Luz Elba used to live with this sister for quite a few years. Her sister was surprised to hear my voice but she was even more surprised when I asked her about her sister, Luz Elba! She told me that her sister moved back to Puerto Rico to live with Papyto and Mamita (her parents' nicknames). I asked her sister, "In what part of Puerto Rico did Luz Elba go to?" She told me, "In Barrazas, Carolina, located in Puerto Rico." Since I'm originally from Vieques Island, which is an island away from the main island of Puerto Rico, I knew very little of how to travel around the many towns of the main island of Puerto Rico. I had no idea where Barrazas, Carolina, Puerto Rico was!

I hung up the phone with Luz Elba's sister. I then told Mom, "Mom, I have to go to Puerto Rico right away! Time is short." Within a few days, I found a flight to San Juan, Puerto Rico, and I booked it immediately!

CHAPTER IX

PUERTO RICO

Upon landing in San Juan, Puerto Rico, I called my grandfather, who lived in Fajardo, Puerto Rico. I knew where Fajardo was because it was where you boarded the ferry that took you to Vieques Island, the island where I was born. I knew I would have to wait almost two hours for my grandfather to reach the airport to pick me up.

When he finally arrived and saw me, we embraced and hugged each other. I love my grandfather. How can I describe him? He has blue eyes, white hair, and he looks like the singer Bing Cosby. He doesn't look Puerto Rican at all. He is a very funny old man. He and my grandmother divorced a long time ago. They had eight children together. He lives by himself. My grandfather is always saying things to women. While we were heading back from the airport, we stopped at an intersection with a red light. While we waited for the light to turn green, we saw this gorgeous woman cross the intersection. My grandfather touched my elbow as to make sure I saw her.

As she walked past our pickup truck, he said, "Hola, Corazon, Tanta's curbas y yo sin frenos!"

Translation: "Hello, Sweetheart, so many curves and I don't have any brakes!" I almost died laughing! That was my grandpa, never a dull moment! We headed back to my grandfather's home. It was late in the evening. I asked him if he would take me to Barrazas, Carolina, tomorrow, which would be Sunday. He said he would, although he was not sure of the area. He said not to worry. We would stop on our way there and asked residents where the Calo family lived.

My grandfather is in his 70s. He loves to engage in long conversations. I'm sure we passed his bedtime hour that night. He was telling me of his young years, at about my age. He was quite a Casanova. He was remembering all his conquests before settling down with my grandmother. While they were married, they had five sons and three daughters. We got up early Sunday morning. My grandfather made me a very delicious breakfast. The eggs came from the chickens he had in the yard. After breakfast, we headed out to Barrazas, Carolina.

Most highways in Puerto Rico are two lanes on each side of the roads. Coming from Fajardo was a single lane on each side of the road leading to the highway, which became two lanes on each side. Driving to Carolina was about an hour's driving time. We came to an exit sign that said "Barrazas, Carolina." I agreed with my grandfather that the road we took, leading us to our final destination of our trip, had to be the most dangerous road we had ever experienced in our lives. First of all, the road was just a little wider than a single road but not by much! This road had so many curves that it should have been named "Snake Road"! There were absolutely no markings on this road warning drivers of the hazardous driving conditions.

For example, there were no white lines separating the opposite traffic coming towards us. Like I said, this road hardly fit two cars going by opposite each other at the same time. In other words, two cars, one going, one coming, they could hardly fit through this road at the same time. One more thing I noticed. This dangerous road was going uphill, so all we had was steep cliffs on either side of the road. There were no barriers on each side of the road, to hold a car from falling down these cliffs. Like I mentioned before, this road had curves after curves going uphill at all times. Oh, and finally, the local drivers drove, in my opinion, just too fast for this road! While my grandfather drove, I noticed that you did not see any cars coming down the opposite side when approaching the curvy road until the car was right there in front of you.

My grandfather was not used to driving on such roads, so when approaching a curve, he would slow down and move to the edge of the curve while making his curve. I noticed that the opposing traffic did the same thing. We also noticed that many female drivers were passing us on the opposite side of the road as well. Locals had no problem driving on this road. We made a few stops along the way, namely business stores, to ask for landmarks, to lead us to where Mr. Sotero Calo resided. Our destination became more difficult to locate. We found out that Mr. Calo's house was not located on the side of the road, like the other houses. It was not even located on top of the hills, like some houses were. His house was located at the bottom of a cliff, away from view of the road! There was a creek running not far from his house.

We made several more stops, asking the locals to point us to Mr. Calo's house. We finally located the location of his house and yes, indeed, his house was not visible from the road! My grandfather and I began descending down this cliff very carefully. We finally saw the house at the bottom of the cliff! Mind you, no one knew us since our arrival was a complete surprise visit. I introduced myself as Luz Elba's friend. Then I introduced by now my exhausted grandfather to them. I immediately asked for Luz Elba and the father said, "She is attending church." Luz Elba was 16 years old and I was 19. She had never mentioned any boys that she was friends with to her parents. So this came as another surprise to them.

Immediately there was a lot of whispering going on among her three brothers and sister. Luz Elba's dad told one of the older brothers to go get Luz Elba and her sister, Zory, who were attending church. Papito, as he was called by everybody in that house, was no more than 5 feet tall. What he lacked in height, he made up with his voice. He spoke very loud and I immediately feared him. Mamita, as she was called by her family, was much taller than Papito. Her voice was normal compared to Papito. Papito spoke to everyone like if we were all hard of hearing. Mamita excused herself, and the next thing I heard was a loud sound of a hen screaming coming from the yard. I asked my grandfather what was going on. He said very calmly, "Somebody is killing a chicken or a hen." He told me, "It definitely wasn't a rooster."

I guess a rooster would sound different if he was being killed. He then said, "It's dinner!" We didn't wait too long before Luz Elba arrived along with her sister, Zory, and her brother, Ninin. Luz Elba immediately asked me, "What are you doing here?" I responded, "I wanted to see you and I also wanted to meet your parents and rest of your family." I then said, "You cut your hair!" Luz Elba asked me, "Why are you here?" Before I answered her, I said, "I want you to meet my grandfather." She shook hands with him. I then said to her, "The reason I'm here is to see you one more time," I told her that I was being deployed to Vietnam for a one-year tour. I also emphasized to her that, as you very well know, not every soldier that goes to a war comes back alive!

Even if you do make it back, some of us return not in the exact condition we went there. Many will return without legs, if lucky, with one leg. The truth is that some soldiers will return without arms, feet, and perhaps without eyes. We just don't really know. We just hope for the best. Luz Elba then began to cry! I stopped then and there and changed the subject. I told Luz Elba I didn't have much time left before I had to report for overseas duty. My departing date was May 4th. I invited Luz Elba and any of her family to accompany me to go to Vieques Island. I wanted to see it one more time. The beautiful island that I was born. I left this island thirteen years ago. I had family there that I would very much like to see.

I asked Mr. Sotero if he had a room for me for a couple of days before we headed out to Vieques Island. If not, I would book a hotel room, but it was quite a distance to the hotel. I would have to rent a car and I really didn't want to drive the road that brought me here. He said loud and clear, "Oh, no, you'll be fine here with us." He said that two sisters would share a bed and two brothers would share another bed so that I would have my own bed. I was to sleep in the boys' bedroom for the duration of my stay here.

We headed for Vieques Island in the 9 A.M. boat. There was the ferry boat, much larger in size that took cars, trucks, as well as passengers. It left at 12 noon. We planned not to stay overnight. We planned to return on the 3:30 P.M. boat. The trip through the ocean just took one hour and ten minutes. The trip was usually suspended during rough seas and bad weather. The boat had an upper deck and a lower deck. Many passengers got seasick, so bags were given out by one of the staff crewmembers. On our way to Vieques Island, we saw dolphins jumping from the water, sort of like putting on a show for us. Spotting dolphins is quite common on this trip to Vieques Island.

At times, the boat hit a choppy wave. I struck a conversation with one of the crewmembers and he said that at some point during the trip, the sea would cross different channels of water and that one could feel the boat react to these channels by the sea becoming choppy at these spots. I completed my one-hour trip at the upper deck. I loved the scenery upstairs. As soon as we left the mainland, we could see, right away, Vieques Island from far away. It was Luz Elba and Mamita's first trip to Vieques Island. Mamita got a little seasick, as most passengers do if it's their first time on a big passenger boat through the ocean.

The boat, although much larger than a yacht, or a fishing boat, but on the other hand, it was smaller than the ferry boat that accommodated cars and trucks as well as passengers. The fare for the trip to Vieques Island was only $2.00 per passenger! For senior citizens it was $1.00 per citizen! However, the fare for the ferry was much higher, if you brought your car or truck. I was also told that one had to make a reservation to travel in the ferry with your vehicle. Right before we arrived at Vieques Island, the boat passed by what was known as Caballo Blanco, translated "White Horse." It is a very small white sandy piece of land. The natives gave it this name. Young, good swimmers swim from Vieques Island to Caballo Blanco.

The distance is considered quite far to swim to it. Right after we passed Caballo Blanco, we arrived at Vieques Island. I rented a car, recommended by my uncle, who lives on the island. Because I was recommended by my uncle, the rental price was only $80.00 for five hours. Vieques Island is only 21 miles long by 4 miles wide. There are about 10,000 habitants living on the island. I have uncles and cousins living on the island that make a living by owning their own business here. They own the only hardware store, the only car wash, one of the few bars, as well as a few grocery stores. One bar is owned by my uncle. The bar is located across the only funeral home on the island. He told me that whenever somebody dies, his business picks up!

I drove Mamita and Luz Elba to one of the most popular sights of Vieques Island: Rompe Ola, translation "Break Waves." Rompe Ola is a road through the ocean. Yes, I said, through the ocean. It was built by the American Marines. Since the Navy and Marines has had a big interest in the island of Vieques, they decided to build a road through the ocean. The Marines and Navy practice their war strategies here at the end of Vieques Island. They built a large port in order for the Navy ships to dock there. Here you would find boats with scuba divers exploring the bottom of the sea, since it is very deep by this port.

This road built, going out to the sea, I would imagine the Americans' goal was to build the road from Vieques Island to reach the mainland. They stopped building the road somewhat ways to the mainland. The builders ran into engineering problems. The ocean became too deep to sustain the boulders that were being placed to continue making the road. The extreme currents that run through these parts of the ocean were too much for the continuation of the road. Now, I'm not an engineer, but why not build a bridge? They built the Golden Gate Bridge in San Francisco, California.
They also built the Verrazano-Narrows Bridge in New York City.

These bridges would reach the mainland from Vieques Island. Perhaps building a bridge was too costly. Anyway, the tourists enjoy driving through this road, admiring the ocean to the right and the ocean to the left. They enjoy seeing choppy waves to their right and calm waves to their left. Now you see why the name Rompe Ola, translation "Break Waves" was given to this road. The road ends at some point in the ocean and then continues with great boulders throughout the path of the road without asphalt, which becomes undriveable. We then drove to Media Luna, translation "Half Moon." This is another popular beach. Media Luna is shaped like a half moon, thus getting its name from its shape. The locals gave its name.

The water is bluish green, with a sandy beach. It's just spectacular to describe it. We wet our feet but because of time, we skipped the swimming part. We continued our tour of the island and drove to another popular tourist place, La Esperanza, translation "Hope." This is a nice part of town located on the west side of the island, along the beach. Many tourists choose La Esperanza to "hang out" during the evening hours. There are restaurants, bars, and of course, the beautiful beach to go swimming.

Mamita, Luz Elba, and I continued our sightseeing. I then took them to El Banario, translation "The Bathing Place." We don't take baths here, we go swimming instead! You see, the Banario is an actual beach, also on the west side of the island. The tourists, as well as the locals, love to go swimming here.

It is one of the most private beaches to go swimming! There are no words to describe this place. It's just beautiful! I met a couple, who were tourists. They confessed to me that in all the times they vacationed in Puerto Rico, had they known about Vieques Island, they would've spent all their vacation in Vieques Island, for sure!

Well, it was time to head back to the mainland. I returned the rental car, which was a walking distance to the boat, taking us back. We boarded the boat by paying our $2.00 each at the gate. Well, I actually paid for my two guests. It was a very emotional moment for me because I actually didn't know if and when I would visit this beautiful Island of Vieques again. I spent my first seven years of my life here and I would miss it. I admired the locals, including my family, because they managed to find a way to make a living here. There are approximately ten thousand habitants sharing this beautiful Island.

We arrived at Fajardo, where the boats dock, and as always the trip took just one hour and ten minutes. It was a beautiful day for traveling through the ocean. Compared to the tranquility of Vieques Island, the mainland is so busy with heavy traffic and busy people doing their thing, whether it is shopping, dining, or doing business with customers. We took a carro publico (taxi) to Carolina. The ride was about an hour. In Carolina, we took another taxi that went to Barrazas, which would take us about another twenty minutes to arrive at Mamita's house.

I spent the rest of the week with Luz Elba's parents and family. During this week I experienced something remarkable. I learned how to swim at the age of 19! Yes, it's never too late to learn how to swim. Luz Elba's brother, who was 14 years old, taught me how to swim. It wasn't in the ocean. It wasn't in the river. It was in a creek that at one point it got to be 6 feet deep. I'm 5 feet 6 inches tall. He showed me a rock or rather a boulder, in the middle of the creek. He swam to that boulder. Then he gave me instructions how to move my legs and hands, thus showing me how to swim to the boulder. Well, I did it! That was how I finally learned how to swim.

The day before I was to leave to New York City, I invited Luz Elba to go with me to "El Pueblo," translation "downtown." Keep in mind that Luz Elba's parents would never allow Luz Elba to go anywhere alone with me. She would have to be accompanied by someone in her family. So, in this case, her oldest sister, Zory, came with us, to go shopping downtown. My idea of going shopping was a surprise to Luz Elba and Zory. We were walking on the sidewalk of the stores. We came to a jewelry store, where I invited them to go in. I then instructed Luz Elba to pick out any engagement ring she liked and I would purchase it for her. They both were shocked, as you can imagine!

Among the many beautiful rings on display, with various prices, she chose one with the price tag in the middle range of the other rings. I said, "If that's the one you like, then that's the one I will buy it for you." She was 16 years old and happy! We headed back to Barrazas but now we were an engaged couple! I felt like I just entered Heaven! When we arrived back to the house, Luz Elba showed the engagement ring to Mamita. She was very surprised and at the same time began to cry and hug her daughter. Then the moment came to show the ring to Papito. Now, remember, I fear this 4-foot5-inch man because what he lacked in height, he made it up in how loud he spoke. He spoke to you so loud that I thought he was very angry with you!

Now, I really didn't know what to expect next. He surprised me by giving me a sermon that day. He said, "You know, this is a serious matter now. This is no game now." Of course, this speech, out of his mouth, was much louder than I was accustomed to hearing from him. All I could say was, "Yes sir, yes sir!" I was so scared of this man that had I been a dog, my tail would be under my butt! Well, the last day was very rough. I hate goodbyes. Saying goodbyes to a family I just met a week ago but already I had bonded with all of them in my heart.

Right before I left, I handed an envelope to each member of Luz Elba's family. I handed an envelope to Papito, Mamita, Zory, Johnny, Junior, Cucusa, and Ninin. These were Luz Elba's family. In those envelopes were my life's savings. I have always saved my money! Remember when I was living in Philadelphia and I mentioned that I had a part-time job, at the age of ten years old, I saved all my money and gave it to my grandmother to save it for me. I also mentioned before that my very first job was working for a major insurance company. I was earning very little money back then. However, that little money, I was able to save it, all of it. Up to now,

I took all my life's savings to Puerto Rico, to meet Luz Elba's family.

Why did I just give my life's savings to Luz Elba's family? I did it because I was on my way to report to a year tour of duty in South Vietnam! I could not predict the future, meaning that I did not know if I would ever come back alive! If, and when, I did make it alive, I did not know in what physical or mental condition I would be. So I had decided to donate all of it to Luz Elba's family. Luz Elba realized that after counting all the money in those envelopes, she could have picked a more expensive engagement ring! After I said my goodbyes to Luz Elba and her family, I headed for San Juan Airport for my flight to John F. Kennedy International Airport in New York.

I arrived at New York after three hours of flight time. The flight was a little bumpy because of the wind turbulence but we made it OK. I took a taxi to Manhattan, to my mom's apartment. Again, after sleeping overnight at my mom's place, it was time to say goodbye one more time. This time, it was going to be goodbye to Mom, my stepdad, my two sisters, and finally, my kid brother. It was hard saying goodbye to all of them. My next destination would be Saigon, Vietnam.

CHAPTER X

TWENTY-TWO-HOUR FLIGHT

When I graduated from AIT in San Antonio, Texas, I received my orders to report to Oakland, California. I was to join my unit there and depart with them by ship. The trip by ship would take about three to four weeks, depending on factors like the weather at the time. However, my orders were changed. My new orders were mailed to me at my mom's New York City address. This was the address on record when I was originally drafted and I received my induction letter from the U.S. Army.

The new orders indicated the change in plans the U.S. Army had for me. My reporting date, May 4th, 1966, was the same date as in the old orders indicated. However, I was not to report to Oakland, California, but instead, it was changed to Fort Dix, New Jersey. There I was to join troops who were headed to Saigon, Vietnam. Our way of transportation was, not by ship, as indicated by the old orders, but now, my new orders indicated that my transportation was by commercial airlines, namely Capital International Airways. The flight would take 22 hours! The captain said over the speakers that we would make two fuel stops; each stop would be done after eight hours of flying time. The first stop would be Anchorage, Alaska. The second stop would be Tokyo, Japan. It would take another six hours of flying time to get to our final destination: Saigon, Vietnam.

For somebody who hates to fly, twenty-two hours in the air is just a little too much for me! I get very nervous and scared, at the same time! While looking around inside the posh plane, full of soldiers, I spotted my best friend, Robin Brown. He was sitting towards the middle of the plane, where the wings are. I convinced the soldier sitting next to Brown to swap his seat with mine. He immediately agreed, since my seat was considered a better seat in regards to feeling the turbulence caused by air pockets that are very common on these flights.

Well, now that I found someone to talk to, I felt more at ease and not think too much of the flight at hand. Brown and I had a lot in common. We were both from the same faith. We were both Christians. We shared the same age. Both of us were 20 years old. Now, I was not surprised to find out that my friend Robin was also a C.O. (Conscientious Objector). We both had fiancés at home. The only difference between us was our height. Brown was a big, I mean really big, 6-foot, 2-inches-tall African-American and I'm a short 5foot-6-inch Puerto Rican!

We just got comfortable and enjoyed the flight to Anchorage, Alaska.

As I mentioned before, we heard the captain announce through the speakers the flight to Alaska would take eight hours to complete. We would then stop for refueling. We were given permission to leave the aircraft, for about an hour, while the ground crew refueled the jet. Brown and I took this opportunity to walk through the airport terminal. After being in the air for eight hours, we needed to do some walking. As Brown and I walked through the terminal, we noticed the Eskimos saying goodbyes to their relatives before they boarded their flights to their destinations. You might say to us, "So what about it?" Well, the Eskimos say their goodbyes very different from us! We hug and kiss our relatives when we say our goodbyes.

The Eskimos, on the other hand, RUB noses, yes, I said, they RUB noses with one another! Now, I asked myself, "Why do they rub their noses with one another instead of kissing and hugging, like we do? Is it because of their culture?" You know what I think? The Eskimos live under subfreezing weather all the time. I think they would frown on kissing and hugging and they would rather RUB their noses with one another! Maybe they think that if they kiss, they will get stuck to each other! That's what I think, too! Have you ever seen somebody's tongue get stuck on a pole because of freezing temperature? Usually you see this stupidity as a result of a DARE between a bunch of idiots and the dumbest falls for it!

Well, our hour was up and we headed for our plane, which was refueled and ready to take us to our next destination: Tokyo, Japan. Since it was another eight hours of flight time to Tokyo, Japan, leaving Anchorage, Alaska, I decided to find a seat where my best friend, Brown, and I would sit near each other and continue our long chat. However, this flight was at nighttime, so we decided to cut the chat short and decided to sleep for most of the flight ahead of us. Again, true to the captain's words, we landed in Tokyo, Japan, in eight hours flat!

As in Anchorage, Alaska, again we were given permission to stretch our feet and walk around the terminal for about an hour. Again, I saw something very peculiar at the terminal. As in Anchorage, Alaska, I observed Alaskans saying their goodbyes, by rubbing their noses among each other, instead of kissing and hugging. Now, in Tokyo, Japan, I observed many Japanese wearing masks over their mouths and noses, you know, like one would see in a surgical operating room! They all looked like surgeons and nurses, walking by the terminal, instead of the operating room. I would expect to see this back home, in a hospital, at the surgery ward, where doctors and nurses, wearing their masks, are ready to report for surgery duties. I had an idea why they were wearing masks, but I had to find out for sure. First I had to find an individual Japanese who spoke English.

It didn't take too long before I found someone who spoke English. I asked him, "Why the mask?" I was hoping he didn't inform me that they all had tuberculosis! No, that was not the reason for wearing the masks. The reason for wearing the masks was to not spread GERMS! This was the same reason I was thinking in my mind. How thoughtful and clever, to all be like that! Well, the hour went by pretty quick, and we headed back towards our plane, which by now was ready for us to come aboard. Brown and I sat together again and braced ourselves for our final destination, Saigon, Vietnam.

CHAPTER XI

SAIGON, VIETNAM

May 4, 1966, I arrived in Saigon, Vietnam. Now wait a minute! Didn't I just fly 22 hours and it was still May 4! Yes, you see, we were now on the other side of our world: Southeast Asia. We were now dealing with the time zone. It was a little confusing, like, we were behind a day here! We would get it back again, when we returned to America, hopefully!

When we landed in Saigon, Vietnam, and the jet made a complete stop, of our final destination, we saw a sergeant come aboard the jet. He took the microphone and gave us instructions to begin exiting the plane, from the rear first. Well, to my surprise, no one moved! The feeling was mutual throughout the entire group of soldiers. We were all scared! Don't you blame us! We were to embark on a one-year journey, to a country we didn't know anything about, who was at war. We were called to assist this country from being taken over by its communist neighbor:
North Vietnam.

The fact of the matter was all plain and simple to all of us on that plane. Some of us would never leave this country alive! For the lucky ones, they would return home with some kind of wound in their bodies. For example, some would lose their legs, or if lucky, only one leg. Others would return back home without their arms, and if lucky, only one arm. Others might lose their hands or fingers or toes. These would be classified as extremely lucky! It's pretty frightening if you think about it. Finally, the sergeant sent two MPs to go to the back of the jet, to begin ordering the soldiers in the last seats to begin moving out towards the front and exiting out of the jet.

DISPOSITION OF MAJOR U.S.
ARMY UNITS IN SOUTH VIETNAM

NOTE: THIS MAP ONLY SHOWS THE LOCA-
TION OF UNITED STATES ARMY TROOPS.
INFORMATION ON THE PLACEMENT OF AIR
FORCE, NAVY AND MARINE PERSONNEL
WAS NOT AVAILABLE.

101

Then the MPs continued to give instructions for the next row in the back to move and follow the soldiers in front of them. This continued until finally the jet was completely empty of all the soldiers that were on board. We were then transported by trucks with benches on the back, to a place called Camp Alpha, located in Saigon. It never ceased to amaze me that everywhere we went, we got to experience something new. The sergeant in charge of Camp Alpha greeted us by saying the following: "Listen up, soldiers, this camp was attacked last month, by mortars." He continued, "So, in the event we get attacked again, while you are here, grab your mattress and wrap it around yourselves, and seek cover wherever you find it!" I told my best friend, Brown, "Thanks for the welcome speech!" I was already coming down with a bad case of DIARRHEA!

We found out that Camp Alpha was a processing camp to cut orders for everyone arriving in country. This was the term used by the staff here of soldiers arriving to Vietnam for the first time. These soldiers arriving to Camp Alpha would be processed and sent to their assigned units as replacement for other soldiers. Now, of course, we were not being told what happened to those soldiers we were replacing. I was imagining that some of these soldiers we were replacing were KIAs (Killed IN ACTION). Perhaps the solders we were replacing had what I heard around here. It is called "THE MILLION-DOLLAR WOUND"! It's a wound obtained while in battle that sends the soldier back home for good!

Now, I didn't know what a million-dollar wound was, but I guess, eventually, I was going to find that out! Then finally, we could be replacing soldiers, very lucky, soldiers that were simply returning home, after their 365 days of service in Vietnam. I have to say that I was impressed with Camp Alpha. I mean, the showers were all lined up with no roofs, divided by walls, and just large enough for one soldier at a time. However, a bunch of soldiers could take a shower at the same time, in their private cubicle shower room. Now, the toilets were nothing like back home, oh, no. In fact, they were not even toilets at all! They were like, in my grandfather's days, like latrines. They had a seat on top of a concrete square big box.

At the bottom was a large drum, which before was a full 30-gallon drum. The drum was cut in half and used as the bottom for the latrine. When you finished doing your business, and the other soldiers, likewise, and the drum filled up with all the business, a Vietnamese papasan removed the drum. He then took the drum someplace else and burned the crap out of it! It was premature but it served the purpose. This was another experience to be mentioned later. On my first night at Camp Alpha, the sergeant came to me and said to me, "Torres, you have guard duty tonight." I said, "Sergeant, I don't do guard duty." He got close to my nose (again) and said, "What did you say, soldier?" I repeated what I just said.

Then I noticed his eyes were getting bigger and closing from time to time. I then said to him, away from his nose, "I'm a C.O." Of course he knew what a C.O. stood for: Consciences Objector. He then said, "Soldier, you're in Vietnam now. This isn't any game here. These are real bullets coming at us at all times." I then told him, "Well, Sergeant, my Lord is real too!" Then he shook his head, left and right, and said, as he walked away, "I just hope I see you here, next year, when your time is up." I was in Camp Alpha for three days. There were no attacks in those three days.

I received my orders.

The orders indicated that I was to report to a unit called the 85th Evacuation Hospital, which was part of the logistical command, located in Qui Nhon, which was 270 miles northeast of Saigon, Vietnam. I was instructed to take a C130 transport plane, which would take me, and Brown, yes, Brown, to Qui Nhon, Vietnam. Before I left Camp Alpha, I asked a soldier, who in my opinion was a very lucky soldier, returning back to the good old USA, in one piece. I asked him if, since he just finished his 365-day tour in country, I wanted to know from him, about Qui Nhon, and the 85th Evacuation Hospital. He said, "Oh, you're going to like it!" Now, I didn't know if he was being sarcastic or he was telling me the truth.

I mean, here we were, rookies, in a combat zone. Is there such a place in a warzone that a soldier would actually say, "I like this place"? When I left for Vietnam, the Rolling Stones just recorded a song, titled "I Got to Get Out of This Place!"

CHAPTER XII

QUI NHON

I arrived at a city of Vietnam called Qui Nhon, after a one-hour flight on a C130 transport plane from Saigon. Qui Nhon is 270 miles North East of Saigon. It is a coastal city since the beach is not far away. I noticed right away, nearby were the Navy ships as we approached the runway at the airfield.

The C130 dropped myself and Brown off in Qui Nhon and continued on its journey, to other destinations, dropping off other replacements, reporting to their assigned units. We had a jeep waiting for us, to take us to our unit, the 85th Evacuation Hospital. I didn't know this but the hospital was actually located at the end of the airfield in Qui Nhon. As the driver was taking us to the hospital, he said, "Look around, soldiers, this is home for the next 365 days!" I realized that all soldiers here counted their days, before returning to America again. To me, all I saw was hills and mountains and jungles all around us.

We were in a valley, next to an airfield. My mind wandered right away, to those hills above us. All I could think of was the VCs (Viet Congs), our enemy, setting up their mortars, to begin shelling us with mortars at nighttime. I did not dare ask the driver, when was the last time Qui Nhon was attacked by the VCs or the North Vietnamese Regular Army? It was better that I didn't know! I had 362 days to find that out! During our short orientation by our lieutenant leader, he informed us that we were medics replacing two other medics that completed their tour of 365 days and were going back to America. I found that piece of information to be GREAT news for the two medics going home as well as for us replacing them. That fact gave me HOPE that with any luck, and GOD willing, my time would also arrive, after 362 more days and completing my tour of duty here!

DISPOSITION OF MAJOR U.S.
ARMY UNITS IN SOUTH VIETNAM

NOTE: THIS MAP ONLY SHOWS THE LOCA-
TION OF UNITED STATES ARMY TROOPS.
INFORMATION ON THE PLACEMENT OF AIR
FORCE, NAVY, AND MARINE PERSONNEL
WAS NOT AVAILABLE.

QUANG TRI
QUANG TRI
101ST AIR CAV DIV
3D BDE 82D ABN DIV
THUA THIER
ASH AU A
DA NANG
1ST CAV DIV (AM)
QUANG NAM
I CORPS
23D INF DIV (AMERICAL DIVISION)
CHU LAI
QUANG TIN
1ST BDE 5TH MECH DIV
DUC PHO
QUANG NGAI
KONTUM
BINH DINH
173D ABN BDE
PLEIKU
PLEIKU
II CORPS
QUI NHON
4TH INF DIV
7/17 AIR CAV REGT
PHU BON
PHU YEN
DARLAC

KHANH HOA
QUANG DUC
TUYEN DUC
I FIELD FORCE VIETNAM
PHUOC LONG
LAM DONG
NINH THUAN
1ST INF DIV
BINH LONG
III CORPS
BINH THUAN
DINH TUONG
TAY NINH
LONG KHANH
BINH TUY
9TH INF DIV
BIEN DUONG
BIEN HOA
GIA
11TH ARMD CAV REGT
HAU NGHIA
KIEN PHONG
KIEN TUONG
PHUOC TUY
CHAU DOC
LONG AN
DINH
AN GIANG
IV CORPS
GO CONG
KIEN HOA
25TH INF DIV
SA DEC
KIEN GIANG
PHONG DINH
VINH BINH
CHUONG THIEN
BA XUYEN
VINH LONG
BAC LIEU
AN XUYEN

109

I also found out that the 85th Evacuation Hospital would be our temporary assignment here. My pants almost fell down when I heard that! Then I heard from the lieutenant that our permanent assignment would be the 67th Evacuation Hospital, also located in Qui Nhon. When I heard this good news, I tightened my belt! As a medic, my job, as I was trained to do, was to attend to the wounded soldiers, or GIs, as we called them. The wounded GIs would be coming from the battlefields, which were cities nearby us. They would arrive in Medivac helicopters. We had a landing field for the choppers to land. Once in a while, I was assigned to go pick up the wounded GIs, straight from the choppers, and put them in a litter. I would need three other medics to help me transport the wounded GI, in a litter, to bring him to our emergency receiving room.

Here, we would attend to them. The medics would change their dressings, which normally would be full with blood. This dressing that the wounded GI would have on him was the dressing the field combat medic first applied to him. In our training, we trained to be field (combat) medics, as well as corpsmen, which was more like my present assignment, working in a hospital environment. You might consider us corpsman, luckier than our comrade field medics. We all had tough jobs, but the field combat medics had a much tougher job to perform because, for obvious reason, he was out there, on the field, in a combat environment.

I was told by my patients that when confronted by the enemy, out there in the fields, and our troops engaged the enemy in fire fights, once a soldier got hit (wounded) by enemy fire, he yelled, "Medic," and the medic ran to his aid while under fire. This was why it was a very tough job for our combat medics to do! Also, while engaged in a fire fight, the enemy focused on the following: The enemy looked for the squad leader, the radio operator, and the medic. Why? Well, the enemy would first try to knock out the leader because then the squad had no leader. The enemy would then try to spot the radio operator because by knocking out the radio operator, the squad had no communication with their backup squads or air support. Finally, by knocking out the medic, the squad had no medical support for the wounded soldiers. You might ask yourself, "How is this possible?"

Many VCs operate from tops of trees. The VCs rely on their snipers to do this job. I was also told, and I could see it, that the average age of the soldiers here was in their early 20s. I myself was 20 years old (at this time). This type of war was brand new to our young troops. They told me they engaged in fire fights with the VCs. Then all of a sudden, the VCs disappeared into nowhere to be seen. Later, our troops found out that the VCs had tunnels built underground where they disappeared and camouflaged the entrance to the tunnels. The troops also found the entrance to the tunnels were booby traps. These were usually explosives or sharp bamboo sticks at the entrance of the tunnels, to cause serious injuries to our troops.

In case these entrances were found, if our troops didn't use caution, it was very easy to lose your life or lose an arm, or any part of your body. The more time we spent fighting the VCs, the better our troops got in engaging and dealing with their tactics. As a medic, I treated all kinds of human injuries. The VCs used the terrain for booby traps. For example, they took a simple bamboo long stick and sharpened the end, so sharp like a needle. At the end of the sharpened bamboo stick, the VCs smeared it with excrete. They then stuck the bamboo stick inside a stream of water, or hid it inside a bushel.

When a soldier stepped on the bamboo stick, sticking up, it penetrated the soldier's lower leg extremity and immediately got infected with excrete on the end of the bamboo stick. The enemy used booby traps in so many ways, that it got back to America. For example, on my time off from taking care of the wounded soldiers, I went to Qui Nhon, downtown, to do some shopping. Though I have to say, I'm not a shopper. I'm a saver of what money I earn. I was getting combat pay, overseas pay, and it was all tax free. This pay was all extra, besides my regular E-3 specialist pay. I decided to go downtown to shop and see what I could find.

Well, I found these cute dolls. The dolls were dressed in the typical Vietnamese dress outfits. So I bought a few to send them to Luz Elba. I also purchased a long chain with a cute beetle for my mom. Well, because of the news in America, about the VCs using booby traps in everything, Luz Elba was afraid of the dolls I sent her. I wrote to her saying that it was OK, since I bought the dolls in a legitimate store and not from the nearby village. What I did buy from the village, which was next to our compound, I bought some good-smelling soup. So did other GIs. Well, we all got very sick to our stomachs. I don't know what got us so sick. I suspect the water the mamasan used to make the soap was contaminated. Why did I suspect the water?

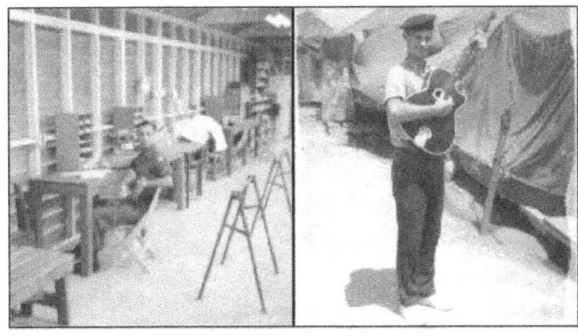

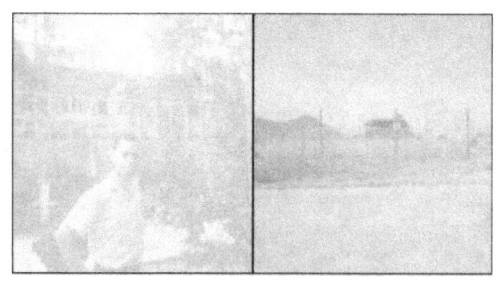

During this same time of the soup incident, I gave mamasan some dirty laundry to do for me. When I received my fatigues (laundry) back from mamasan, nicely pressed, I put them on. It didn't take too long before I got this tremendous itch around my groan area. Yes, I gave my underwear to mamasan to wash for me. I developed a red rash around my groan area. I ended up going on sick call for it. I told mamasan, "No more laundry for you. Your water is no. 10! Our water is no. 1." We compared everything here in Vietnam, it was either no good (no. 10) or very good (no. 1)!

If you said to papasan that he is no. 10, he would take that as an insult. He would respond back to you that he was not a VC (Viet Cong), our enemy. He would tell you that he was not no. 10, he was no. 1! Not too long after this, one of our GIs caught two papasans making a drawing of our campgrounds. These were Vietnamese that the Army hired to do details around the camp, but supervised by GI. The maps were confiscated immediately and the two papasans were turned over to the MPs for interrogation. I would assume these two papasans were friendly by daytime and VCs by nighttime. You never knew what activities surrounded us.

On the subject of activities, there wasn't much to do around base camp on our time off. So I and two buddies of mine decided to head out to the hills, even though we knew it was off limits. We were dressed in civvies (civilian clothes). You know when you get so curious, you just got to find out. Well, we convinced ourselves to go and check it out! You might call us STUPID and yes, I now agreed, we were very STUPID, but we never took any stupid pills that morning! We headed out to the hills, through the jungles. We were not the Three Musketeers, but rather the three STUPID STOOGES. What were we looking for? You would know better than we did! Well, guess what! We did find something very incredible!

Halfway up the hill, which we knew was off limits to us, we found something. We came across some buildings. These buildings were beautiful. They were painted in beautiful colors. We came across a nun. Yes, a nun, in the middle of the jungle! She told us she was French but she spoke English. We asked her, "What is this beautiful place here?" She said, "This is a leprosarium!" Then she asked us if we were lost! We said no but I felt like saying, "We are just stupid!" She then asked us if we would like a tour of this place. I looked at MOE and LARRY and said, "Sure."

She took us to many parts of several buildings. I asked SISTER MARY LUCY if she would allow me to take picture. You see, when we got back to our base camp, nobody would believe us, nobody! So I took pictures. I took pictures of Vietnamese with their deceased of leprosy. Some were missing fingers, hands, toes, feet, ears, and other parts of their bodies. It's a terrible disease! I found out from SISTER MARY LUCY that whatever occupation these leopards had before they became sick, they continued to practice their occupation here. If they were farmers back home, then they raised rice in the rice paddies and so forth and so on.

Leprosy is a contagious disease and this was why these leopards had to be kept segregated. I asked Sister Mary Lucy if she was afraid of being contaminated. She said no. Sister Mary Lucy was like a missionary, along with others in the staff, who volunteered to do this kind of charitable work. Before we said our goodbyes, I asked Sister Mary Lucy, "Do you get any visitors?" She said, "Only VCs, once in a while." My pants almost fell to the floor! I should have never asked her that! Now my curiosity went to a different level and that was not all, we had to head back to our base camp. I was not looking to meet anybody on our way back! Well, guess what? You are so right! We met somebody.

We met somebody we didn't expect to meet out here in the jungle!

On our way back to base camp, we met a platoon from the Korean forces. They belonged to the Tiger Division. They were patrolling the hills, looking for the enemy, but instead, they found us! Now, once we met with them, some of them began to whisper among them and began to laugh. Now, I know that you know what they were thinking. Yes! Yes! These Koreans were thinking, "What in Hell are these three IDIOTS doing here?" I felt like saying to them, "We are here on a search-but-not-destroy mission!" I didn't say that because they already figured it out that we must be the most STUPID human beings on the planet EARTH!

I approached their leader and said, "We just came from visiting the leprosarium!" He said, "What leprosarium?" I told him, "You should check it out! You might even find some VCs!" Before we said our goodbyes to these outstanding Korean soldiers, I said, outstanding, because I heard somebody mention the Korean soldiers were helping us to fight this war. The difference between us and the Koreans was the Geneva Convention. You see, when we captured the enemy, we fed them, and we even took care of their wounds. Yes, I took care of a wounded VC while two MPs guarded him. The Koreans, on the other hand, captured the VCs, then they interrogated them. Once the interrogation was over, the VCs were of no use to them, so they shot them.

This was why I also heard that the VCs would rather fight us anytime than to fight the Koreans. Just before we said our goodbyes to our friendly Korean soldiers, I saw a Korean soldier with an XK49 grenade launcher. I never had any training with any weapons because of my C.O. status. However, I was very curious about this weapon. Not in the sense to cause harm to anybody. Oh, no, I would never do that! I was just curious about the weapon itself! So I approached him and I almost died because of his garlic breath! I kept my distance and I asked him, "Could I fire your XK49 grenade launcher?" He asked his superior officer for permission, and his officer said yes.

Well, he showed me how to aim and fire this weapon. So, I took the AK49 grenade launcher, with the grenade locked in place, and aimed for the top of a large tree, not too far from where I was standing. I pulled the trigger and after a jolt to my right shoulder, the grenade headed for the top of the tree. When the grenade hit the top of the tree, it exploded. When it exploded, to my surprise, I heard a SQEAKING, crying sound of something. Whatever the grenade hit and exploded at, it fell down to the ground. I saw pieces of blown-up squirrels in the ground! I felt so sorry for those squirrels. I gave the AK49 grenade launcher to the Korean soldier, who was smiling ear to ear! He then pointed to the blown-up squirrels and said to me, "No. 1, chop chop!"

We headed back to base camp and we reached base camp with no other incidents. Of course, nobody believed us, until I took my roll of film to be developed downtown. When I picked up my developed pictures, I showed them all the proof, by my pictures, of everything that just happened to us on that day. While going downtown to pick up my pictures, I saw a black dog crossing the street. I began to smile, for no reason at all, until I heard a babysan (Vietnamese child) say to me, "Hey, GI, that

No 1. chop chop!" Now, I just realized that up to now, I hardly saw any dogs around! I have to confess, I love all animals, even horses that bite, like the horse that bit me a long time ago, but I will never forget that.

Now I found out that this country ate dogs for dinner! This was too much for me. Anyway, I was glad to be back at base camp, or I thought I was going to be glad! Among us medics was a soldier named Robert. Now Robert, to me, was a very interesting human being. For one, he was always complaining about working with us, you know, the hospital duties and all. He desperately wanted to be in the action, as he called it. He wanted to be a frontline combat medic. He told us that he put in a 1049 several times. A 1049 is a form requesting to be transferred to another unit. He had been denied both times, on grounds that he was needed here at the hospital. He was an excellent medic.

His frustration amplified tenfold when he received a "Dear John" letter from his girlfriend back home! Now let me tell you that Robert was Irish, and yes, he liked to drink on his time off. My problem with Robert was that his girlfriend was Puerto Rican. I'm Puerto Rican! Robert showed up in our tent with his "Dear John" letter in his left hand, and a long, sharp knife, in his right hand. He looked directly at me with his dark blue eyes and told me his girlfriend dumped him! There were other GIs with us, but he only looked at me! I almost crapped in my pants! I didn't know what his intensions were with that knife in his hand.

I began to search with my eyes to see what I could grab quickly, should he decide to throw this knife at me. If he decided not to throw the knife at me, maybe he decided to come charging at me with it. Either way, I was getting prepared for some action in that moment. Don't worry, he didn't take it out on me. We advised him that something better awaited him back home. He told us he was going downtown and that he was going to deal with it there. We suggested for him to leave the knife with us. I'm assuming he was going downtown to get stone drunk! I didn't see Robert after that "scary" day. You see, I got transferred to my permanent unit: the 67th Evacuation Hospital. It is still located in Qui Nhon.

From what I was told, the 85th Evacuation Hospital was being phased out and the 67th Evacuation Hospital was taking over its medical duties. I and Brown were part of its new staff of medics coming from the States to join in the replacement of the medical staff. I was assigned to the new orthopedic ward. There were 33 beds assigned under my care. At first I made a point to get to know my patients on a personal level. One day we received a soldier with a gunshot wound on his right instep foot. This soldier was a big African-American, who could easily be a football linebacker back home. I asked him how he came about to getting wounded in his instep.

I found it very interesting when he got shot. Well, he said he accidently shot his foot while cleaning his rifle. Well, yes, I found that to be very interesting, but it gets better. As I took care of this wound on a daily basis, it seemed to this soldier that the wound was healing too rapidly. This soldier began to worry that this wound was healing too rapidly, and that he would be sent back to his unit, to resume fighting the enemy in the jungle. He almost begged me to skip a few treatments and not to do such a thorough job treating the wound. I felt so much compassion for him but I just had to do my job. Not too much later, he was dismissed from the hospital and a medivac chopper took him back to his unit.

Like many of my patients, I never saw him again, which could be a good thing or a bad thing. In either case, I will never know. Out of all my patients, one sticks out. McGuire was a patient full of energy, hopping up and down the ward, always smiling, blue eyes and blond hair, and very tall. By the way he carried on, you would never know the extent of his wound, until you saw it for yourself. His whole front, from just below his throat to his belly, was held together by a thick wire, sawed together. He acquired this wound by exploding scrap metal, hitting his frontal body. He was the youngest of two brothers and two sisters.

He had a very nice-looking girlfriend waiting for him when he got to go home again. He had been in country now for 192 days. On this note, every GI knew exactly how many days they had completed in Vietnam. We all started with 365 days and counted backwards, until we reached one, the day we went back home again. McGuire went to follow-up surgery on the day I was off. When I returned back on duty, I asked for McGuire. I was informed that he never made it out of surgery. I was devastated! I took the news pretty hard! So hard I took this devastating news that I promised myself not to ever get personal with my patients again. From this day on, I would tend to my patients' wounds and nothing else. However, done saying that, a patient came to me with a special request!

Jose was a conscientious soldier. He told me about being out in the bush (jungle), as he called it. Please note, I did not asked for any details. Jose asked me if I had any men's perfume. I was curious at this request. I said, "As a matter of fact, I do. I have a bottle of CANOE (men's perfume) back at my tent." He asked me, "Could you bring it with you next time you come back on duty again?" I said yes. It seemed to Jose that being in the jungle for the amount of time spent there, you began to smell. He felt he still smelled, even though he was now recovering in the hospital.

As promised to Jose, I arrived the next day with my bottle of CANOE perfume. I gave the bottle to Jose. He immediately splashed my CANOE all over his entire body! He then gave me back what little amount left of my CANOE! I said to Jose, "You can keep it." You want to see a HAPPY face! Never mind his wounds, he simply forgot about them, momentarily. He never stopped thanking me enough! On another occasion, we received a Green Beret soldier, who was wounded by a VC sniper. Luckily, he wasn't captured. I truly do admire the Green Berets. I remember them taking medical training with us at Fort Sam Houston, San Antonio, Texas. They did it all and they did it by themselves. While I was in Vietnam, there was a song written about the Green Berets. They were special and that was why they were called the Special Forces.

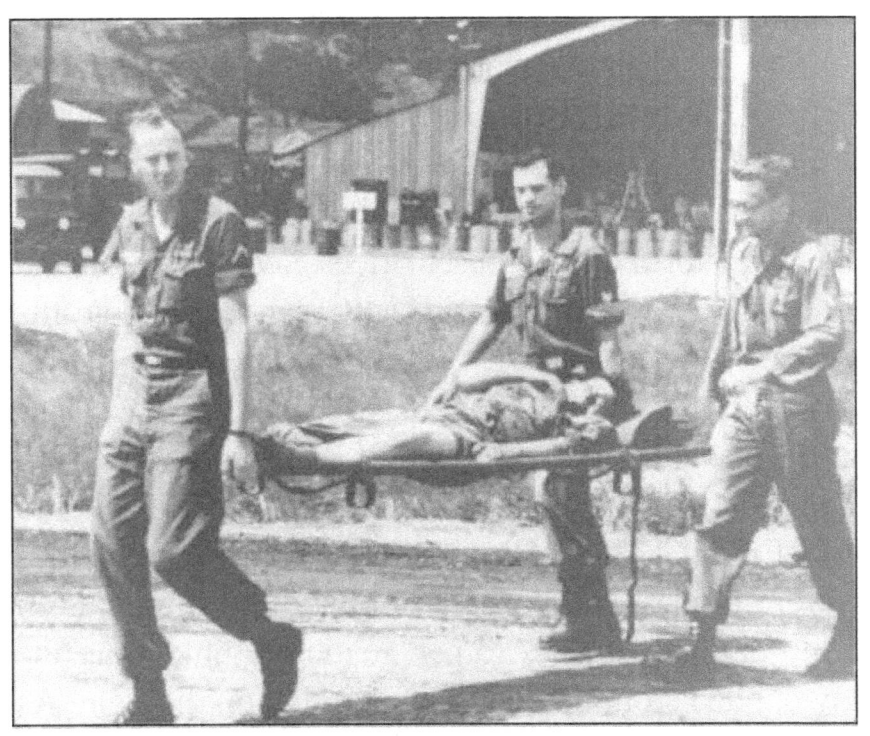

So mamasan came by to do some cleaning around the ward. The Green Beret began to speak to her in fluent Vietnamese. She began to smile from ear to ear. As she smiled, she showed her black teeth, fully showing through her lips. I tell you, that made my day, for sure! The Green Beret soldiers are incredible! On the subject of Green Berets, there was an officer of the North Vietnamese Army (our enemy) that was captured. In his interrogation, he spoke about a Green Beret that parachuted from a C130 and landed on top of a tall tree in the jungle.

Unfortunately, he landed in a spot where a company of NV regulars were around him. The enemy spotted him right away. The Green Beret began to shout to them, in Vietnamese, from the top of the tree. He said, "Surrender now, before it's too late."

The North Vietnamese officer gave the command to his men to shoot down the Green Beret from the tree. This was how the Green Beret died. The officer said to his interrogator, "I wish my men were as brave as that Green Beret!" On occasions when U.S. troops were engaged in battles near us, like in PLEIKU, PHU BON, or BINH DINH, casualties were higher than normal. I sometimes got a call on my day off to assist other medics in unloading the wounded from the medivacs landing on the field. As I approached the medical chopper, I saw one of the most GORGEOUS women on earth. She was either a nurse or a medic. I never had a chance to ask her.

My job was to get the wounded soldiers out of the chopper as fast as possible to our ER facility. When we were done unloading the wounded soldiers from the chopper, it took off to its next destination. I never saw her again! When doing this assignment of picking up the wounded from the choppers, we worked in teams. Each team consisted of three medics and a litter. A litter was made out of canvas material, held by two wooden poles. One pole on the right side and the other pole on the left side, with the canvas in the middle, where the wounded lay down, to be transported to our ER room. The tallest medic would take the front of the litter by himself. I would take the back right side and the other medic would take the back left side of the litter.

On this occasion, my team faced a dilemma. It seemed that the medic on my back left side of the litter had one of his legs, left leg, shorter than his right leg. When we picked up the wounded soldier and began to rush him back to ER, the litter, along with the wounded soldier, began to bounce up and down on the left side of the litter. It was very uncomfortable for the wounded soldier on his way to the ER. I couldn't yell at the medic. It wasn't his fault for being here with a short leg. I blamed the Army for allowing this to happen. The Army had to be very desperate to allow this individual to be inducted into the U.S. Army. I was amazed at what I saw here!

The 67th Evacuation Hospital consisted of various staff members. We had medical doctors, who were part of the surgery team. We also had medical corpsmen, like me. Last, but not least, we had our nurses. I have to say, we all did a great job, considering the environment we found ourselves to be in. I had 33 patients in my ward under my care. I made my rounds, read the patients' charts, and gave them their prescribed medications, at the proper times, as instructed by the medical physician. For the patients who were to be given a daily shot consisting of a needle, a syringe full of medication, these patients made my day. Why do I say that? Well, I developed a technique of giving shots that my patients complimented me on my technique.

The patients compared the nurse's technique versus my technique and I won by a long shot! My technique for giving a shot was: I instructed my patient to lie on his stomach side. I brought his pants down. Then I SMACKED his buttocks, YES, smacked his buttocks. I then proceeded to inject my patient, his prescribed dosage of medication, V E R Y S L O W L Y until done. The patients related to me that they never felt anything! They only felt when I slapped their buttocks. There was no way they felt the injection, none at all! I observed the nurse's technique. She instructed the patient to drop his pants and lie on his stomach.

She then injected the medication into his buttocks and pushed the syringe hard until done with medication. The patient and I agreed. It was very painful, indeed. On the other hand, my technique was painless! As I said before, I followed the doctor's orders written on the patient's chart, which was in front of his bed. However, as human beings that we are, we sometimes make mistakes. I don't know how, but as careful as we are, we make mistakes. I went by the chart, matching the bed number, as well as the patient, for that bed. I approached bed #12 and said to the patient, "Ready for your meds!" He said, "Yes, Doc." Patients called us medics "Doc." I handed over to him two 100ml of Darvan pills.

He looked at me, kind of surprised. Nonetheless, he swallowed the pills, along with the glass of water I gave him. I continued my rounds of issuing the medications. When I arrived at bed #21, I said to the patient, "Ready for your shot?" This patient gave me the surprised look! I instructed him to drop his pants and lie on his stomach. Once he did, I slapped his buttocks and injected him his CCs of penicillin. The next day I reported for duty, as usual. The head nurse called me into her office. She informed me that patient in bed #12 was relieved that he wasn't getting any more shots, according to the patient's remark to her.

On the other hand, patient in bed #21 was complaining to the head nurse why all of a sudden he was being administered injections of penicillin, instead of his usual Darvan painkiller pills! As you can see and I can't explain how in the world I made the mistake of giving out the wrong medications to bed #12 and bed #21. The head nurse chewed me out pretty bad. The patients were fine. No complications occurred, especially with bed #21, getting Penicillin injection shots. I was so glad this patient wasn't allergic to Penicillin; otherwise it could have been pretty bad for the patient and for me too.

After reaching 280 days of being "in country," I decided that I was ready for some R&R (Rest & Recuperation). The Army will grant you R&R for one week to any place you choose. Well, I always wanted to go to Honolulu, Hawaii, and I knew a pastor living there, and would like to spend my week there with him and his family. At this time, he wrote to me that his wife got very sick and he had to fly her back to Texas, to attend her medical needs. So I chose Plan B. For my second choice, I had a friend who was a medic stationed in Japan. He wrote back to me and said he would be happy to see me! He was single and obviously, no wife.

Since I now received confirmation from my buddy in Japan to go and see him, I submitted my forms for R&R to Japan, for one week. I couldn't wait to get out of Vietnam, even if it was only one week. On top of that, I would be visiting another Asian country.

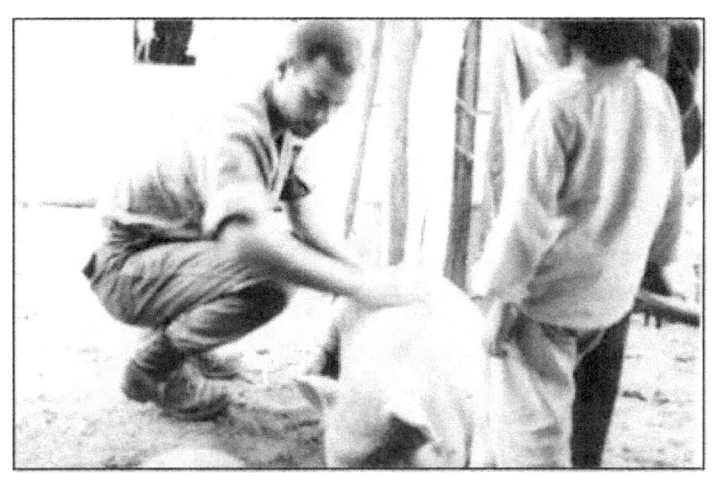

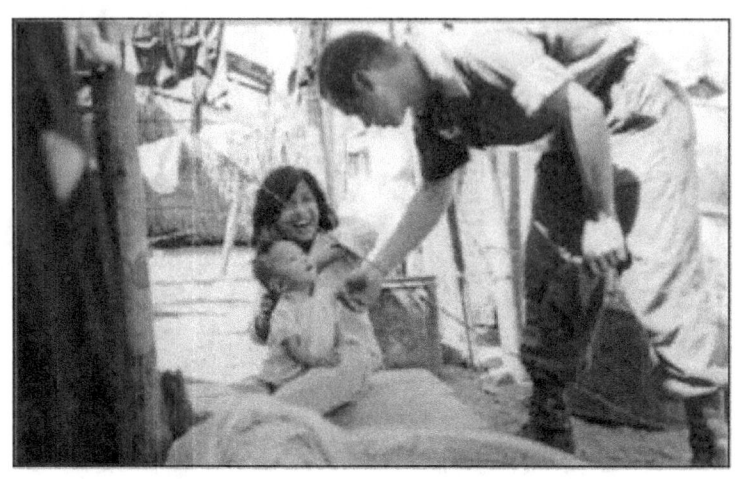

CHAPTER XIII

JAPAN

I landed in Tokyo, Japan's airport after just completing eight months in Vietnam. I knew I'd be spending just one week here but I did need the time off.

My buddy was stationed in Tachikawa, Japan. So I was looking for the train that went to that city. Before searching for that train, I had to convert $400.00 into Japanese money called yens. The Japanese would not accept any other currency. Also, I would have to spend all of the $400.00. The Japanese would not convert any yens left from my $400.00 back into American dollars when I returned back to Vietnam.

There were so many trains leaving for so many cities. All the destination signs were in the Japanese language. I had to stop and ask an elderly Japanese gentleman to direct me to the train that went to Tachikawa. I gathered he didn't speak any English, but he did understand when I mentioned the city of Tachikawa. He directed me to the right train, but he then did something else. He extended his hand so as to get paid for his services. I took out a couple of yen bills and gave them to him. He bowed his head and left with a big smile on his face. I immediately wondered if I over-tipped him!

When I boarded the train, I noticed quite a lot of Japanese passengers were wearing masks covering their mouths and noses. I felt out of place because number one, I was not wearing any mask, and second, I was a foreigner. These Japanese in masks reminded me of the surgeons at the hospital in Qui Nhon, Vietnam. I came here to temporarily forget about my hospital duties and here I was reminded of what I just left! I arrived at my buddy's base camp and got off the train. Then I saw something I'll never forget. The streets were extremely busy with cars speeding on their way to their destination. I did not see any lights in the intersection I was on.

All of a sudden, I saw an old Japanese lady begin to cross the busy intersection. Did you know that the traffic all came to a sudden halt and the lady crossed the intersection until she got to the other side? As soon as she was done crossing the busy intersection, the cars took off again, speeding away. To me, it appeared like something you would see on a TV cartoon channel. I was so surprised and, at the same time, so impressed by what I just saw! Well, I had to ask a Japanese man near me who witnessed what I just saw. I asked him, using sign language. I looked at him and raised my shoulders. Then he answered me back by pointing to the old Japanese lady who just finished crossing the street. He raised his index finger out of his right hand.

Then he pointed to the speeding cars on the road. He then raised his right hand again and this time he showed me two fingers. I acknowledged by bowing my head up and down. What I gathered from all of this is the following: The pedestrians in Japan come first. The cars come after the pedestrians. Therefore, since the pedestrians always come first, they always have the right of way! Oh, and by the way, I was surprised that the Japanese man didn't charge me for that interesting information!

I spent my days with Jose, my friend. While talking to him, I found out that when he finishes his time with the Army, which would be one more year, out of three years he signed up, he was going back to Puerto Rico. He was going back to school. I asked him, "What are you going to study?" Since Jose was a medic for the U.S. Army, he wanted to pursue a career along these lines. He then told me he wanted to be a mortician! He wanted to be an undertaker for a funeral parlor, perhaps, and eventually a funeral director. Now I knew why he could not keep a relationship for too long. Once his girlfriend found out what he wanted to do in his life, they all ran away!

I'm sure they were thinking that Jose was not going to touch them after he had been touching cadavers all day long! Jose also told me that he wanted to move to a big city, like Santurce, Puerto Rico. I asked him, "Why Santurce?" He said that Santurce is a big city; therefore, big cities have large populations. He continued, "Where there are large populations, there are also large amounts of deaths." According to my friend Jose, this would be very profitable for him! On my last day in Tachikawa with Jose, I asked him to take me shopping. We came to a big, giant store. It appeared to me like a Macy's department store. I told Jose that I wanted to visit the electronics department.

We took the elevator to the 12th floor. I told Jose that I had to spend the rest of my money that I had converted into yens and that I was not allowed to convert what was left into American dollars. When we entered the 12th floor of this big department store, I was very surprised at the items' prices. To me, they were all bargains and not too many on sale. I went to see binoculars being displayed on the shelves. There, I saw these huge binoculars that really caught my eyes. They were the biggest on the shelves, as well as the biggest I had ever seen. Then there was the price. I could not believe it.

The price was only $160.00. I just could not resist purchasing them. I wanted to take them back with me to Vietnam. I could see myself looking to the hills, back in Qui Nhon, with these beautiful binoculars. In those hills, when I saw white smoke. Who was there?

What were they doing up there? Were they cooking up there? Now I had my high-power binoculars to check it out! I couldn't wait! I had plenty of yens left, to purchase these lovely binoculars. They were a BEAUTY! I also saw some tape recorders at fabulous prices but they were too big for me to carry back to Vietnam. I could have purchase it and shipped it back by mail to the States but to me it was a big bother.

I did buy a nice portable radio. It was black and had gold trimmings around it. I also visited the jewelry department. There I purchased some 12k gold chains, which were at awesome prices. My week finally came to an end with my friend Jose. I wished him luck with his future plans as a mortician. I wished him to be successful and for us to continue to be in touch. I considered my one-week stay in Japan to have been very rewarding and educational at the same time.

Now I had to head back to Vietnam to complete my last 120 days in the country.

CHAPTER XIV

120 DAYS

I arrived at Cam Ram Bay, Vietnam, after my six-hour flight from Tokyo, Japan. They say, "It's a small world." Well, you might ask me why I say that.

While I was making my transfer arrangement from the jet I just got off to the C130 plane that would take me to Qui Nhon, Vietnam, I saw someone at the airport that looked very familiar. It was my cousin, Noel. Yes, Cousin Noel lived 10,000 miles away on the island of Vieques, Puerto Rico. I asked him, "What are you doing here?" He said, "I'm delivering U.S. Mail to the airport." The mail got sorted here and got distributed throughout the different cities of Vietnam. I wouldn't be a bit surprised if the C130 I was about to take now, there might be some mail going to Qui Nhon right now.

I said my goodbye to my cousin, Noel. I told him to take good care of himself while in Vietnam. I told him, with some luck we would make it out of this place and perhaps meet up again in Vieques Island, Puerto Rico, where we had so much love. The flight from Cam Ram Bay to Qui Nhon on the C130 took only one hour to complete without any incidents. I saw all my buddies back at base camp and said hello to all of them. I then saw Robert. You know Robert. He was the GI who received that "Dear John" letter from his Puerto Rican girlfriend.

Robert asked me what I bought in Japan. I told him I bought a few gold chains, a portable radio, and high-powered binoculars. As a matter of fact, I wanted to check out the hills that surrounded our base camp. I always believed that our base camp might someday be attacked by the VCs. I thought they would attack us during the night, when we were the most vulnerable. I brought my eyes close to the binoculars and what a sight! The binoculars were as powerful as I anticipated they would be. Robert asked me to let him see what I was seeing. He said, "WOW." Everything looked so close to us from way back in the hills. Then Robert shocked me!

Robert fell in love with the binoculars. He wanted to buy them there and now. He told me he needed these binoculars more than me. He said, "You know I'm a lifer." If you never heard the phrase "lifer" and what it means, well, I'll tell you. We call a soldier a lifer if he intends to make a career in the military. A career in the military consists of 20 years or more of consecutive years of active duty. Now Robert knew that I had already mentioned that I was drafted into the U.S. Army. I would do just my two years' active duty, my two years' active reserve duty, and finally my two years' inactive reserve. And that was all I was required to do and no more and no less after that.

Robert convinced me to sell him my precious binoculars. I sold him the binoculars for $160.00! Yes, $160.00, no more and no less. After I made the sale, I began to think about this transaction. Boy, was I STUPID! Have you ever heard supply and demand! The supply was only one but the demand was very high! I could have sold to Robert my binoculars for TWICE the amount I paid for them. He would've definitely paid the price! I guess you know by now I'm not much of a business guy. I didn't show Robert my gold necklaces or my radio. I'm afraid I would have made some more bad deals with them should Robert have shown some interest in them.

My days in Vietnam were getting shorter and shorter. It became nerve wracking. When a soldier here reached 180 days to go home, you just couldn't wait until it finally happened. I was now down to 100 days for me to go home. I was now considered a short-timer by the rest of the troops around me. When you became this close to going home, you didn't want anything to go wrong. You didn't want any surprises. Well, it happened! We had a visitor. He was a general. He paid us a visit to Qui Nhon. He was looking for medics. Lucky for us, we had our own, General Myers, who was our regional commander. General Myers informed that other general that his medics were staying right here and weren't going anywhere, but to their assigned duties, here at the hospital.

Not even Roberts was leaving Qui Nhon! General Myers understood very well how much in need that other general needed medics for field duty. He was just going to get his medics from Saigon, just like we all processed there as replacements. I'm so grateful for General Myers holding on to us, especially me, who now had become a lucky short-timer.

Out of all my experiences in Vietnam, one of many stands out. We had doctors and nurses on our great staff, besides us corpsmen. These doctors and nurses were officers serving in the military. Doctors had a rank of captains. Nurses had a rank of either 2nd or 1st Lieutenants.

The enlisted men, like us medics, were not permitted to socialize with the officers. It was just how it was. We had a nurse on our staff. She was absolutely gorgeous. She met a helicopter pilot, here in Vietnam. They fell in love. He proposed to her. Her engagement was short and sweet. I'll explain later. We had another nurse on our staff. She met a doctor, who was also on our staff. They fell in love. The doctor also proposed to her. She also said yes. They got engaged. Her engagement wasn't for long either! You're probably wondering, what was going on with these couple of officers?

Okay, here was what was happening. These engaged loving couples decided to get married in Vietnam. Yes, here in this country, in Qui Nhon, where the North Vietnamese were fighting the South Vietnamese, to rule, and the good USA was in the middle of it! Well, you would have never guessed it. War or no war, these two couples were getting married today. Now, I didn't know if this double wedding would be picked up by the press. It was a fabulously story, if you ask me. The wedding had everything that you would expect in the States. It had its bridegroom, its best man, all in tuxedo, and its bridesmaids all in gorgeous gowns. There were flowers all over the place.

For the ring bearers, the couples chose these cute Vietnamese children, two pairs of boys and girls. They too were dressed to the T. They were spreading flowers down the aisle of the church. We had invited a chaplain who performed the beautiful ceremony. The couples answered their vows when it was their turn. It was all done in duplicate form. It was simply a day to remember and I believe it was history in the making. They took a week off and went on R&R. To the Vietnamese present and watching there, this is how it's done back home! I was winding down my tour of one year in Vietnam. However, battles were being fought all around us, places like Pleku and Ankay.

Forces fighting the enemy were the 1st Cav. Division, the 25th Infantry Division, and the 101st Airborne Division. We took care of all their medical needs. From these units I met soldiers that I trained with back in basic training in Fort Sam Houston, San Antonio, Texas. Some of these soldiers decided to sign up for airborne training. After basic training, they continued their training and went to jump school to get their wings. Once their training was completed, they were assigned to the 101st Airborne Division. They were now here in Vietnam, fighting the enemy, or Charlie. Somebody decided to name our enemy Charlie! I don't know why this name was given to identify our enemy.

Whoever named our enemy Charlie didn't like Charlie! Unfortunately, I now met some of these soldiers I took basic training with, not under favorable conditions. I got to meet them, all shot up. It was my job to help them and tend to their wounds. Some of the wounded soldiers said to me, "Torres, you got it made here!" I just nodded my head. I held back from saying, "I didn't volunteer for anything other than what I was assigned." When I was in basic training, one day an officer helicopter pilot gave us a visit. He pitched his speech on anyone interested in becoming a chopper pilot.

Some of our trainees showed some interest in expanding their training to become helicopter pilots.

Helicopter pilots get shot down along with their helicopters here in Vietnam all the time. The pilots that did survive ended up here in our hospital. Again, I passed up helicopter flight training school and gave up obtaining my wings. For now, these wings were broken from these pilots and airborne soldiers while they recuperated from their wounds. Upon graduation from airborne and flight training, you got awarded wings! I just focused on medical training, and that was all!

Well, I'd spent 170 days here in Vietnam since arriving from Tokyo, Japan, from my one week of R&R. I was down to the last ten days. It was hard to believe. As for us getting attacked or shelled by the enemy, nothing doing. You might say, how in the world can that be? There was definitely a war going on in this country. I even heard the sounds of war around us all during the night! You asked any troop, whose base was somewhere else, and he would tell you that they were attacked on such and such a day. What about their air fields? When was the last time their air field got mortared or shelled by the enemy? Again, their response would be on such and such a day!

Now, in turn, if they were to ask me those same questions, I would respond, "I've been here 355 days and we haven't been attacked yet, not our campsite, our air field, or our hospital." Like I said before, all around our neighboring cities, battles were being fought constantly. Why not us! Well, for one, the Korean Tiger Division patrolled these hills surrounding Qui Nhon. The enemy would rather not engage with the Koreans and their war tactics. Second and most important for me personally was my LORD. I had so many Christians back home in New York City, praying for my safe return home.

I really did believe that my Almighty GOD had not permitted the enemy to attack the city of Qui Nhon yet. I will keep you posted once I leave Qui Nhon for good. Today I not only just completed my last 170 days in Vietnam, I also completed my last ten days in Qui Nhon. It was May 4, 1967, and I couldn't believe I was boarding a C130 en route to Cam Ram Bay. It would take approximately one hour to get there. I met other GIs from other Vietnam cities that were also rotating back to the States and going home! We arrived at Cam Ram Bay, one-hour flight, with no incidents. I say this because there had been incidents where C130s had received enemy fire from the grounds!

We had to wait a little while until we were given the OK to board the jet that would take us back to the good USA! The time finally arrived when we were given the OK to board the jet. It was nice to see American flight attendants again on this flight. All the GIs were deeply TAN from spending a year in

Vietnam, although six months of the year was the monsoon season, where it rained every day. The stewards complimented us on our TAN faces! The atmosphere on this jet was inexplicable. I'm having a hard time describing it. I mean, here we were, soldiers who just survived one year (365 days) in a combat zone, going home at last!

I'm remembering that a year ago, arriving in a jet similar to this one, landing in Cam Ram Bay. When the jet stopped and the sergeant in charge said to all of us soldiers to begin exiting out starting from the rear of the jet. What happened? No one moved; remember that! No one wanted to leave the jet, no one! Well, now it was the total opposite! We could hardly wait to board this jet. Every GI on this jet was extremely excited, happy, joyful, and cheerful, indeed! Once the jet was airborne, leaving the air field, I got the biggest surprise of my life! It seemed the soldiers were holding back their emotions until this moment.

The majority of the GIs began to holler profanities and obscenities at the country they were just flying over (VIETNAM). The four-letter word was being shouted by the majority of soldiers on this jet, followed by the word VIETNAM! My virgin ears had no choice but to hear all of it. Once in the air and having left Vietnam, our flight captain announced over the loud speakers, greeting us and thanking us for our services. By this time our GIs had calmed down to a more civilized mood. The captain also announced that we were heading for Tacoma, Washington. There, we would make our connecting flights to whatever city we were heading for. In my case, New York City.

However, before we made our connecting flights, we were told to follow instructions once we had landed in Tacoma, Washington. What instructions, we all wondered. After we finally reached our destination, Tacoma, Washington, we waited in the jet for further instructions.

CHAPTER XV

TACOMA, WASHINGTON

When the jet stopped at its assigned gate in Tacoma, Washington, a sergeant from the airport boarded the plane.

He greeted us and welcomed us back to the good USA. We all cheered at once. The surprise was that we were all going to be treated to a steak dinner, reserved for us at a restaurant located at the airport. No one was to leave until they had their steak dinner. The closest thing to steak in Vietnam was dark buffalo meat that I asked mamasan to get me from time to time. I must admit, it wasn't bad at all. The restaurant was nice and fancy. We had all forgotten how life is in the good old USA. Why, just yesterday, we all feared for our lives in a foreign country, and now we found ourselves safe and sound, back in the USA!

We ordered our steaks to our preference, whether it was well done, medium rare, or raw. I ordered mine well cooked. We had a choice of baked potatoes, French fries, or sweet potatoes. I ordered sweet potatoes. I also ordered sweet corn on the cob, what the hey, why not? They also brought us bread with sweet butter. Not bad for our first American meal! For my drink, I ordered a large virgin piña colada. I don't drink anything with liquor. For my dessert, I ordered some good old-fashioned apple pie. It had been a whole year since I had a meal like this one!

After dinner, I and my best friend, Robin Brown, said our goodbyes. You know, Robin Brown, who I first met a year and a half ago, at the induction center. We had been together for basic training, AIT, Vietnam, and now we were saying goodbye. You know, the Army, according to the recruiters, two buddies can enlist together. The two buddies can elect to stay together for the entire three-year service, without getting separated. This was exactly what happened with me and my best friend, Robin Brown. The only difference was that we did not enlist under the buddy plan. We were both drafted into the U.S. Army with no plan at all.

It turned out great for us and we didn't need a recruiter to join us for three years under the buddy plan. Robin Brown was heading for North Carolina, for his 30-day leave. I, on the other hand, was heading for New York City, for my 30-day leave.

CHAPTER XVI

30-DAY LEAVE

We arrived at Kennedy International Airport after a seven-and-a-half-hour flight from Tacoma, Washington.

It felt so good to be back home again. One gets to appreciate everything around you tenfold. Things that one takes for granted, now that one has a second chance, one cannot imagine how much more they appreciate it. A ride on the subway might be just normal for most New Yorkers who commute this way, but now for me, not so much. You see, for me riding the subway was now a new experience that I now cherished. I had now been given a second chance to enjoy riding on the subway. I had not taken a ride in the subway for the last year and a half.

A ride that I took for granted before, now I thanked my LORD, he had permitted me to once more ride the subway. When you spend one year in a war-torn country, where life can be taken from you in a split second, where you are denied a way of life that one is accustomed to, one can appreciate things that we took for granted. Yes, we all could appreciate one more time a subway ride in New York City. I visited my church for Sunday services. I sat down towards the middle of the bench, towards the middle sitting area. My pastor, who was sitting at the pulpit area, must have spotted me.

When it was time for him to preach, he acknowledged my presence and asked me to speak to the congregation. When speaking in public, I'm one of very few words. However, this time I took the opportunity to truly thank everyone who spent time praying for me and all the soldiers that were there serving time in Vietnam. I thanked everyone who prayed for our safe return home. I'm one testimony of those prayers that had been answered. I truly thanked them for that. I never realized it but all the young youths of my church said that I looked good! Some even said that I was in good shape physically! The girls liked my TAN look in my face and arms.

I told them that in Vietnam, we had six months of consecutive rain and the other six months we had nothing but hot, dry weather, and absolutely no rain. That, I told them, was how we soldiers acquired this TAN. I was glad to be home again. My mom was so glad to see me, as you can imagine. She was so happy to see me again, safe and sound. The happiest person to see me again was my fiancé, Luz Elba. Of course, she was one of the most faithful girls, among everyone else that was praying for my safe return home again. We went out and spent some time together, but of course, we had a chaperone with us at all times. Luz Elba was living with her oldest sister, husband, and their three children.

Luz did a lot of babysitting and took care of many things around the house. She also knows how to cook too! All these quality things I like about Luz, besides her physical look, that all attract her to me! I'm just glad she gave me a chance. I say this because I had a "Playboy" status around the youth church members. I thought I never would have a chance with her. I expressed some interest in Luz when I attended her 9th-grade graduation. At that time, her oldest sister, whom Luz was living with, her sister flatly came out and asked me if I was romantically interested in her sister. I said yes.

When I was drafted into the U.S. Army, Luz, unbeknown to me, left her sister and moved back to Puerto Rico, to continue living with her parents again. She continued her high school studies in Puerto Rico for the next two years. After that, she returned to the Bronx, New York, to live with her sister again. She completed her senior year at Roosevelt High School, where she graduated. Luz began to work and we resumed our dating. We enjoyed our chaperone dating until my 30-day leave ended. I was to report to my next assignment. I was assigned to Fort Hood, Texas. This would be my last assignment in the U.S. Army.

CHAPTER XVII

FORT HOOD, TEXAS

What a coincidence! My two years of active service would consist of total time served: one year in Texas and one year in Vietnam.

I remember being asked once before I completed my AIT, the officer in charge asked me, "Where would you like to be assigned stateside?" I said, "Fort Dix, New Jersey." He asked me, "Why?" I said, "Because it is the closest state to New York City." Then he asked me, "Where would you like to be stationed overseas?" I said, "Puerto Rico or the Caribbean." He again asked me, "WHY?" I said, "Because either Puerto Rico or the Caribbean, both are nice and hot!" Well, I don't know why he asked me for my choices of assignment. I guess I was sent to Vietnam because it was hot over there! As for being assigned to Texas, it is kind of hot, too! Both assignments were way off from what I selected.

I believed the officer representing the U.S. Army had no intentions of fulfilling my choices of serving stateside or overseas side. So here I was in Fort Hood, Texas. The first thing that impressed me from the get-go was the armored tanks. There were rows and rows of tanks. It's too bad that these tanks were here and not in Vietnam. The terrain in Vietnam was not suitable for tanks, as in other wars, in other countries. I mean, there were tanks in Nam but their use was limited. I believe Fort Hood, Texas, was where the 1st Armory Division's base is. The Division did a great job during World War I and World War II. It's a big base and I like it a lot.

Someone told me that Fort Hood was where Elvis Presley was assigned to before being assigned to Germany, for his overseas tour. They say there were girls all over the place because of Elvis being here. I'm a big fan of Elvis because his style of music is unique. I felt good in Fort Hood. For one, this would be my last six months of active duty in the U.S. Army. As a Vietnam veteran, that now I was, we were assigned a red patch to wear on our side of our uniform or fatigues. For those troops that hadn't been to Nam as of yet, when we passed by them and they noticed the red patch on our shoulders, they looked at us from the corner of their eyes.

I wondered what they were thinking. They probably envied us because we had already been there and made it back, alive! While they, on the other hand, were getting ready to go there.

While I was in Fort Hood, I decided to take the road test for a military license again. I say again because before going to Vietnam, I took the road test and I failed it. I was so disappointed because I didn't pass it. You know what? My Lord was in on this! Yes! You see, in Vietnam, the Viet Cong put mines on the roads so when a military vehicle ran over a mine planted on the road, the military vehicle blew up, causing major injuries to the soldiers in those vehicles. I took care of some of these drivers in Vietnam that were wounded in this fashion.

You see, most of the Vietnamese used bicycles as a means of transportation and not vehicles. So say I had obtained my military license before I went to Vietnam. In Vietnam, I could have been given a medical vehicle, like an ambulance, to transport patients back and forth. There would be a high probability that my vehicle would also drive through a mine and blow us up. So my Lord, again, was watching over me. He decided that even though I wanted a military license, my Lord was not going to permit it. So, while I was in Fort Hood, I decided to take the road test for a military license again. Guess what? I passed it! Mind you, it was the same test I took before going to Vietnam. This time I passed it.

It was a piece of cake! You know what? There were no mines on the roads of Fort Hood, Texas. Now that I had my military license, I would get assignments to drive military officers around base camp. One day I was assigned to drive 2nd Lieutenant Rodriguez to the other side of Fort Hood, Texas. Now, since day one, when I first met Lt. Rodriguez, he never seemed to hit it off with me. I don't know the reason, but I sensed he didn't enjoy my company. Maybe it was the first time he engaged in a short conversation with me. He asked me, "How was it in Vietnam?" I said, "Sir, it's nothing like here."

In Vietnam, everybody had a gun (except C.O.s). "Everybody respects everybody. Soldiers depend on each other for their survival. Trusts are more there than rank." He then said to me that an officer was there to lead. I said, "Yes, sir, you are so right. That is why on a search-and-destroy mission, the officer, in this case, you, Lieutenant, will be the squad's point man." He got all shook up when I said this to him! Well, I drove Lt. Rodriguez around in my military jeep. I don't know what made me drive my jeep like I did. For instance, I was taking sharp turns where the jeep's tires on one side were lifting from the road. I would come to a screeching halt on a red traffic light. Then when the light turned green, I would speed up and shift my gears as fast as I could.

At one time I glanced at the lieutenant's eyes and it scared me. They looked like they wanted to come out of his eye sockets! When we arrived back at headquarters, the lieutenant jumped out of my jeep and entered the building. He shouted at the company clerk, who was sitting at his desk, "Don't you ever assign Torres as my driver, ever again. Is that understood, soldier?" He then turned around and called me "Parnelli Jones!" I said to the company clerk, "Who is Parnelli Jones?" He told me, "He is some famous SPEED RACER."

On another occasion, I was returning my duce and a half to our motor pool. A duce and a half is a truck with, in my case, there were benches on the back of the bed truck. In this type of truck, we transported soldiers to various areas around base camp. I came to a stop sign and made my stop. The minute I began crossing the intersection, out of nowhere came this red Corvette and slammed to the side of my truck. There was no damage to my military truck; however, the red Corvette sustained quite a bit of damages. On the front side, it looked like an accordion. Immediately, a lieutenant jumped out of his Corvette. I stood at attention and saluted him, like I was supposed to do when an officer was in your presence.

He then said to me, "Don't you salute me, soldier." He seemed to be very upset! So I responded to his comment, "OK." We waited for the MPs to arrive. The MPs arrived and asked me, "What happened?" I gave them my report. Then they asked the lieutenant, "What happened?" I don't know what the lieutenant told them. I was told to continue driving to the motor pool. As for the Corvette, it was being towed away to somewhere. I asked the lieutenant if he needed a ride. He just stared at me and moved his head slowly, back and forward. I took that for a no.

I informed the sergeant in charge of the motor pool about my accident. He asked me to show him where the truck was hit. I showed him and he said, "There is no show of damages." I said, "I know." So I hung up the truck keys and put them with tomorrow's schedule of events. We were scheduled to go to the field for our field exercises. Now, I flatly told my sergeant that I didn't need to go on this field exercise. I reminded him that I just got back from Vietnam, and that I didn't need this training. He would not excuse me no matter what I said. When we returned from the field exercise, we were told that tomorrow we had to attend a ceremony. We were told to wear our full green dress uniforms for this occasion.

The occasion was for Specialist 4 Army Medic Flores. He was being awarded the Bronze Star for Valor. He showed exceptional bravery while under a fire fight. He responded to the cry of "medic" to the wounded soldiers, going to their aid, numerous times, while under fire. Well, tomorrow came and we were all ready to begin marching outside to the parade. However, we had a problem. Specialist Flores was MIA (Missing in Action). We found him in the bathroom. He was so drunk he could hardly stand up! I said to him, "Man, you better get it together, this is your day!" He said, slurring his words, "I don't give a frog (I substituted the word he actually used) about no medal!" We splashed some cold water on his face and helped him to get fully dressed. We practically dragged him to formation.

He was able to receive his Bronze Star. I just don't how he was able to stay in attention and not pass out! It was an unbelievable day. Right before my ETS (Expire TIME SERVICE) came up, I received my paperwork for the medals I earned while on active duty. They were the following:

- Vietnam Campaign Medal
- Vietnam Service Medal
- Vietnam Gallantry Cross Medal
- National Defense Service Medal
- Good Conduct Medal

My ETS (Expiration Time Service) finally arrived on November 17, 1967. This was my last day in the United States Army. It was a HAPPY day, indeed! The two years of active duty in the U.S. Army had been an unbelievable experience for me, to say the least. It was a happy day for me.

However, two years ago, other men were drafted on the same day I got drafted. Unfortunately some of those men were sent to Vietnam, like me, but never came back alive. Perhaps others did come back, but not in the same physical condition. Regrettably, some came back without an arm, or a foot, or an eye, or without sight. Then there are, most of us, who will have to live with the unforgettable memory of the war we just finished experiencing. There is a term, the military informed us, referring to war memories. The military calls it PTSD (Post-Traumatic Stress Disorder). On November 17, 1967, I said my final goodbye to my best friend, Robin Brown. It's hard to believe that Brown spent two years with me, minus the 60-day leave, minus 7 days for R&R = 663 days together! We both got drafted on the same day. I met him on that day.

We were assigned to Fort Jackson, South Carolina, for our processing. The Army sent us to Fort Sam Houston, San Antonio, Texas, for our basic and AIT. For overseas duty, we were both assigned to Qui Nhon, Vietnam. We were also assigned to the same unit in Qui Nhon: the 67th Evacuation Hospital. After our tour in Vietnam, we both were assigned to Fort Hood, Texas. I tell you, this very rarely happens. The odds were stacked against us on these same assignments being assigned to us for two years. I said my goodbye to Mr. Brown.

He told me he was going to South Carolina. I told him I was heading back to NYC, Manhattan, again!

CHAPTER XVIII

MANHATTAN

I got drafted in Manhattan, NYC, two years ago and now I was back home again in Manhattan, NYC. NYC, the city that never sleeps. Like I said, a year and a half ago, while I was being interviewed on television, when asked, "Where are you from, soldier?" I said, "New York City."

Then the interviewer asked me, "And how is NYC during this time of Christmas?" I responded, "Swinging, man, swinging!" Everybody laughed, including the staff behind the TV cameras! I came back to live with my mom again. Due to the fact that I was contributing financially to the household when I got drafted, I applied for financial help to the Army, to help my mom financially. The U.S. Army obliged and began sending her money on monthly bases while I served in active duty.

While I'm on the subject of the U.S. Army, I was watching the news on television and Chad Huntley just reported that the city of Qui Nhon was just attacked by the North Vietnamese Army. The 365 days that I spent in Qui Nhon, the city never was ever attacked while I was there. This is no coincidence.

My LORD did not permit this city to be attacked while I was there, and that's a fact!

A lot had happened in the two years I was away from home. I am the oldest of four siblings in my family. First, I found out that my older sister got married and joined her husband in Germany. My younger sister also got married and moved out to an apartment, with her husband. My kid brother also moved out and went on his own. My mom got divorced and now had a new boyfriend. My mom began having mental issues as soon as we switched churches. My stepdad looks like Rock Hudson, the actor. Women began to perspire the moment they saw him.

The pastor's wife played guitar in church. I always wanted to be a musician. I will elaborate more on this subject at a later time. I asked her if she would teach me how to play the guitar. I said, "I'll come to church for lessons." She said, "Oh, no, I'll come to your place!" I later found out that she was interested in my stepdad! My stepdad and I joined the choir at the church we decided to attend. I also found out that the choir director, she was also interested in my stepdad.

Now that I was back and eligible to work again, as far as employment was concerned, because I was drafted, my employer was obligated, by law, to hire me back. However, the position I held when I was drafted at my workplace did not have to be the same position when I did return to my employer.

When I was drafted into the U.S. Army, I was employed by a major insurance company in New York City. At the time, I was hired to work in the mailroom. I was also attending college during the evenings. While I made my rounds, delivering mail to different departments (more on this subject coming up later), I always stopped at this particularly long hallway. In this hallway was large, long window panels. I would stop and observe these guys, dressed in suits and ties, working in the computer room. I said to myself, "Someday I would like to do that kind of work!" Remember this, more on it later!

When I reported back to work at the insurance company, I went to the personnel department. The personnel department assigned me to the new policy department. My new job title was now "Certificate Control Clerk." I set some goals for myself, in regards to my employment career. Even when I was drafted into the U.S. Army, I was attending part-time college during my free evenings. Now that I was employed again, I was again attending college during the evenings. At the time I began attending college, my goal was to become a CPA (Certified Public Accountant). I enjoy math. I like working with numbers. I graduated from high school.

On graduation day, I was honored to receive the top award in the math department. Now that I was employed again, I was resuming taking college courses again, something happened in the last two years. I began to hear about these IBM accounting machines being used everywhere. I then said to myself, "Why I should pursue a career in accounting when these machines will probably take over the accountants' jobs?" So I changed my mind in becoming a CPA. I focused on studying IBM. Since now I was a military veteran, I was eligible for benefits. While in Vietnam, I said to myself, "If and when I get out of this place, I'm going after all the benefits I am entitled to."

One important benefit that I just found out was that I did not have to be in the two-year active reserve. I was excused. I also found out that I was also excused from the two years of inactive reserves! In the past, all military personnel were obligated to serve a total of six years total, namely two years' active, two years' active reserves, and two years' inactive reserves. I was now eligible for education benefits under the G.I. Bill. I could now enroll in an educational institution and get educational financial benefits at the same time. My new goal now was to pursue a career in IBM instead of becoming a CPA. So to do that I had to change my curriculum of studies, leaving accounting classes and switching to IBM courses.

I enrolled in a trade school located in the Empire STATE Building in Manhattan, NYC. The name of the school was the School for Computer Studies. I was on my way to work with IBM Computers, HURRAY! I graduated from computer studies. At this time, I was still working at the insurance company as a certificate control clerk. I wasn't making much money as a clerk. In fact, my kid brother dropped out of high school, and he was making more money than me. He was working as an auto mechanic.

I approached my boss and requested a transfer to our electronic processing department, since now I had the necessary courses completed in IBM. The transfer was granted. Remember when I was working in the mail room, delivering mail to different departments, I would stop by this large window panel in the large hallway to see these guys working in the computer department, well, here I was. I reached my goal. I was now working side by side with the guys I used to look at from the outside panel window. I was now working with offline computers.

However, I then began to hear that the real money to be made was in Wall Street, Lower Manhattan, and NYC. So after working at the insurance company for four years (credited of two years while in the military), I decided to try for an opportunity at working for a brokerage firm in Wall Street. I would be doing the same line of work, running offline computers but in another company. Hey, guess what? I was interviewed by a well-known brokerage firm in Wall Street. Part of the interview consisted of wiring a board for an IBM reproducer to reproduce IBM cards. Of course, I knew how to do that! We were instructed at school how to wire boards.

When I was transferred to the insurance electronic processing department, I was already wiring boards. My interview went very well and I was hired! I gave my two weeks' notice at the insurance company. I was now going to make a little bit more money than my kid brother. HURRAY! HURRAY! That HURRAY quickly turn out to HURRY, HURRY! I'll explain. Not too long after I began working for the brokerage firm, in which I was operating their computers, on the nightshift, a fire broke out! I worked the nightshift all by myself. I worked on the 7th floor. The fire began on the 7th floor. It began on the other side of the hallway, away from the computer room. It spread very rapidly from the 7th floor and proceeded to the 8th floor, and spread onto the 9th floor.

The building security guard came running to the 7th floor and knocked on the window of the computer room and signaled me to open the computer door. I opened the door and he yelled, "HURRY, HURRY, you have to evacuate the building immediately!" As I followed him down the stairs, I noticed flames coming down from the stairs going to the 8th floor. There was smoke in the stairs. We hopped down the stairs and out of the building. We were now safe. I called my boss, who was at home. I informed him about the fire in the building. He then asked me, "Did you do a shutdown of the computer?" I said, "NO!" He then asked me, "Did you put away the data disks in the safe?" I said, "NO!" He then informed me that our safe was FIREPROOF! He then said that he was on his way.

As it turned out, we got VERY lucky. The fire did not penetrate our computer room. Smoke did enter the computer room but did not damage our data files on our numerous disks. I later found out how this fire started and spread so fast through the 7th, 8th, and 9th floors. The maintenance crew was stripping the old wax from the hallway in the 7th floor. To do this stripping, they spread this liquid over the floor to strip the old wax. This liquid loosened the old wax on the floor. To remove the wax from the floor, they used a machine called a BUFFER. The liquid they used was extremely flammable. Well, at one point, the operator of the buffer swung the buffer too fast, side by side. The buffer hit the corner of the hallway, and a spark came out of the buffer.

As you can imagine, ALL HELL broke loose! The wax remover liquid (called STRIPPER) turned into flames immediately. The fire immediately spread to the 8th floor, feeding on all the paper in the offices, and continued onto the 9th floor. The fire took place on a Friday night. On Monday morning I was told we had our first fire drill! Now that I had been operating offline computers for the past two years, I decided to go back to school, to advance my career in computers. I wanted to learn how to operate ONLINE computers. Just like an offline computer, known as tab operators, do everything offline, computer operators do everything online.

Computer operators mount tapes to tape drives, mount disks to disk drives, and most importantly computer operators give system commands to computers, to perform different tasks in a matter of microseconds! The computer operator has more responsibility than the tab operator. That was why the computer operator got paid a higher salary than a tab operator. So these were the reasons I wanted to go to school to become a certified computer operator. Three days before completing my course on becoming a certified computer operator, I made an announcement to my classmates. I informed them that I would be transferring to the ONLINE computer operations department of the firm I'm presently working for.

Let me explain. This was a huge step for me. Even though you take the course and graduate, it is not easy to get a job running online computers without any experience. Most firms are asking at least six months' experience before they trust you in running their online computers. I made friends with the online computer manager of the firm I work for. He is Peruvian and I'm Puerto Rican and we speak the same Spanish language. I informed him that I would be graduating from online computer study course in three days and that I would like to transfer to his department. He told me that at this time, he did have an opening; however, the position was on the third shift at nighttime. He also mentioned that the pay was the lowest of all computer operators.

However, because it was the third shift, it was compensated with a 15% nightshift differential rate pay. I said, "Mr. Gambol, I'll definitely take it!" I said to myself, "Regardless of the starting salary, it is definitely more than what I'm presently being paid!" So I worked in the Online Computer Operations Dept., 3rd shift, for two years and there were no fires for that amount of time. After working for this brokerage firm for a total of four years, an opportunity came up to work for an investment bank. It was offering a lucrative retirement plan, 401(k), and great starting salary. However, the hours of working for this bank was also the nightshift. Since I had been working nightshift for a while now, I decided to go for an interview.

The interview went great, again! The manager agreed that I had enough experience with running online computers to offer me the job after interviewing other applicants. I came home, very excited indeed, and told my mom the good news! I said, "Mom, you are not going to believe this! I'm going to work at THE WORLD TRADE CENTER!" Upon giving her my excited news, her face turned into a very sad and worried face. She then told me, "Son, be very careful working at THE WORLD TRADE CENTER." She told me that she had a premonition that something bad was going to happen to that building. I said, "Oh, Mom, nothing is going to happen!" She said it to me in June 1970!

I worked at the World Trade Center for approximately ten years for the investment bank. After these ten years, the bank decided to build their own building in Princeton, New Jersey.

CHAPTER XIX

NEW JERSEY

The bank had a dilemma in dealing with its staff members. The bank wanted to keep their staff but in actuality, they didn't know how many staff members were willing to commute to Princeton, New Jersey. To compensate the staff employees, the bank was offering $300.00 a month for three years for commuting expenses to all employees who were willing to stay and continue working for the bank. The commuting package was a great deal for me. You see, by this time, I had gotten married to my fiancé, Luz Elba! We got married just two years after I finished with my active duty in the U.S. Army.

I was able to get married in two years after active military duty because I had always saved my money. While in active duty, serving in Vietnam, I was getting overseas pay, combat pay, specialist 4 rank pays, and the best part was that all this pay was tax free! I did not gamble my money away. I definitely did not give my money to the prostitutes who were waiting for you downtown! I did not SHACK UP with the pretty young Vietnamese girls who were charging $1,200 a month, like some GIs were paying. Oh, no, I saved my money! What I did was send my savings passbook, along with a military post order check (it's like a money order), to my bank in Manhattan, NYC, every month.

The bank, once it received my deposit, would send me back my passbook with my deposit credit to my account on it. So this went on for twelve months while in Vietnam. So we got married after I ended my two years of active service and went on our honeymoon to Mount Airy Lodge in the Pocono Mountains in Pennsylvania. We spent one week in the mountains, in the month of June. The food was out of this world. After three days, I went for a massage. The masseur was an elderly oriental gentleman. Now normally I would be very skeptical about this situation. I said to myself, "Should I feel uncomfortable at any time, I'm walking out, with clothes on or no clothes on."

P. O. BOX 419969
KANSAS CITY.
MISSOURI 64141
(816) 921-2200

DEAR PROSPECT:

THE KANSAS CITY ROYALS BASEBALL TEAM IS HAVING TRYOUTS/WORKOUTS AT COLUMBIA
UNIVERSITY AND WOULD LIKE YOU TO ATTEND.

COLUMBIA UNIVERSITY IS LOCATED AT 216th STREET AND BROADWAY IN NEW YORK CITY
(BAKERS FIELD). PLEASE BRING YOUR EQUIPMENT AND BE READY TO PARTICIPATE. A PEN
OR PENCIL IS ALSO NEEDED TO FILL OUT YOUR PLAYER INFORMATION CARD.

PLEASE APPEAR AT THE TIME AND DATE LISTED BELOW. GOOD LUCK AND SEE YOU THERE!

Thursday - June 29, 1989 at 08:30 A.M.

P.S. IF YOU NEED TO CONTACT ME, MY PHONE NUMBER IS:

(718) 967 - 3842

Sincerely,

Ike Cohen
Ike Cohen

186

188

Let me tell you, this masseur was all professional. He restored my energy level back to 100%. He massaged all the right muscles with this nice smelly oil that did wonders for me. I needed this massage due to the activity that was scheduled for us later that day. The lodge set up a baseball game between our waiters and the bridegrooms. I couldn't wait to play baseball again. I batted two hits in four at-bats. One hit was to left field batting right. The other hit was to the right field batting from my left side. As for fielding, our manager asked us where we felt comfortable fielding. I told him I would like to play in the outfield.

I requested right field. In the seventh inning, a left handed hitter hit a deep line drive to me, which went over my head. He rounded to first, continued on to second, while I pursued the ball heading for the fence. The ball hit the fence and ricocheted back to me. The runner, to my surprise, continued running to third base. I then, to my amazement, threw a one hopper to our third baseman, who tagged out the runner for the third out of the inning. When I arrived at the dugout, one of the players on our team asked me, "Have you ever played in the pros?" I said, "No, but I would love to play professional baseball."

We lost the game 4 to 2. That was the bad news. The good news was that now that I was married, I found an apartment in the Bronx, New York. Luz and I settled in a nice three-bedroom apartment in the Grand Concourse area of the Bronx, NY. Since I continued to work nights, after six months of being married, Luz suggested we start a family. Our son was born nine months later. We love our son to death! Not too long after our son was born, we were in the Cross Bronx Expressway, when we were rear ended by a car behind us. My son fell to the floor of the front seat. There were no restrictions for child seats. Luz was very upset, as you can imagine.

The other driver and I exchanged IDs. It was heavy traffic. Both of our vehicles were drivable. So we drove away and agreed to report the accident to our auto insurance company. It was my first auto accident. When I reported my accident to our auto insurance company and gave them the other driver's information, I received a big surprise. The other driver's license belonged to a driver that is now deceased! There was no one living in the address of the name listed on the license. So the bottom line is that I ended up paying for the repairs of my own car. I learned a big lesson from this accident!

Our son was only a few months old when Luz and I decided to move to Staten Island, New York. We found a duplex apartment in Staten Island. This duplex apartment had two bedrooms upstairs. My son was upstairs sleeping. When he woke up, Luz asked me to go bring him downstairs. I went to get him and started to go down the stairs. He was inside a bassinet. I carried the bassinet with my son in it. The stairs were wooden stairs. I was wearing socks. I slipped on the top stairs and came down the stairs, hitting each stair with my buttocks. I never let go of my son on my way down the stairs!

Luz decided to find a job and go to work. We had to find a babysitter. We were recommended, from a mutual friend, who is a member of the church we go to, for the ideal babysitter. We hired the babysitter to take care of our son. She was the right person and we loved her. My son loved her, too. When our son turned one year old, our babysitter wanted to know if we would be interested in renting the upstairs of their house. This would be very convenient for all of us, as far as babysitting was concerned. It would be ideal for everyone because we wouldn't have to drop off our son to the babysitter anymore, especially in the cold weather.

We would be renting five rooms on the second floor, plus three more rooms in the attic, should we need to. The rent was very reasonable. We took the offer and moved in. We lived in Staten Island for three years. My son was three years old when my daughter, Becky, was born. When I lived in Staten

Island, New York, I took the Staten Island Ferry to go to work at the World Trade Center in Lower Manhattan, NYC. The fair was a nickel, yes, only 5 cents for one-way fare! The problem with the Staten Island Ferry was their schedule. The last Ferry leaving Manhattan was 12:30 midnight to Staten Island. My nightshift ended at midnight, 12 A.M. One day I was running late to catch the last ferry to Staten Island.

I ran to the terminal. I decided to continue running to the lower deck, where the cars loaded up to the ferry. The ferry had just started to depart from the dock. I never stopped running. I jumped to the ferry and made it to the boat. Had I not completed that jump successfully, I would have been turned into chopped liver by the propellers of the ferry located on the back bottom of the boat!

Becky was only three weeks old when Luz and I decided to move to New Jersey. I used my G.I. Bill to purchase a house in Jackson, New Jersey. I continued working for the bank, even though I now lived in New Jersey.

I became a commuter. It now took me two hours to get to work from New Jersey to the World Trade Center in New York City. The best way to commute is by bus, like so many of us who worked in NYC did. At this same time, the computer staff elected to work three days a week on twelve-hour shifts, Mondays, Tuesdays, and Wednesdays, for shift A. Shift B would also work twelve hours on Thursdays, Fridays, and either Saturdays or Sundays. To be fair to all, shift A would work their days for three months and then rotate to shift B. Shift B would work for three months and then would work shift a days.

On this note, there was a shift A and shift B handling days only. Shift A and B days would not alternate with the nightshifts. If any operator was called in to work for somebody in the other shift, it would be an automatic twelve hours of overtime. The twelve-hour shifts were great for us because of the time off we had during the week. However, for me it was very hard because I commuted four hours daily, plus twelve hours of shift work gave me sixteen hours away from home. That left me only eight hours a day to do whatever! Like I said, the good news was that the grueling shift was only three days a week.

I remember one payday Friday, I took the bus as usual at the bus terminal. I took my seat and closed my eyes, as usual, for my quick two-hour nap. The bus headed out of New York for New Jersey. Halfway of the trip, I woke up and looked out the window. I looked and the terrain did not look familiar to me at all. I got up from my seat and walked to the bus driver and asked him, "Where are you going?" He said, "Atlantic City, New Jersey." I was expecting he would say, "Lakewood, New Jersey." You guessed it, I took the wrong bus. I got on the express bus to Atlantic City, New Jersey, by mistake!

When I arrived in Atlantic City, I immediately took the next express bus to New York City. However, I explained to the bus driver heading to NYC what just happened to me tonight. I asked him for a big favor. I requested for the bus driver to leave me at the Lakewood exit, while he was on the Garden State Expressway. I knew this was an express bus to NYC, and I was asking for this favor. He obliged and I was very grateful and couldn't thank him enough. The bus driver came to the Lakewood exit on the Garden State Parkway, pulled over the shoulder road, and let me out. I remember it was pouring rain. I called my wife to pick me up at the Lakewood exit. My wife hates to drive in the rain, so I knew I was going to hear it from her.

She picked me up after waiting for a half-hour. Yes, I was right; on our way home, still raining, I had to hear it all! Not only that, she did not believe me on taking the wrong bus to Atlantic City on a payday Friday! Because I chose to continue working for the bank, I also was entitled to the commuting package of $300.00 a month for 36 months. The commuting package offered by the bank that I worked for was such a sweet deal for me. Now that the bank relocated to Princeton, New Jersey, I needed to upgrade my transportation. I owned and drove a 1966 Chevy Impala, which had over 100,000 miles.

Since the bank was already giving me $300.00 a month towards my hourly commuting to work, I figured I'd visit a close-by dealership near my town of Jackson, New Jersey. I walked into a Pontiac dealership and a salesman greeted me at the front entrance. I told him I was interested in a sports car. At this time there was a popular TV show called *The Night Rider.* Well, I was fascinated with the car on that TV series. The car on the TV series is a Pontiac Trans Am. I told the salesman to show me a Trans Am in the show room. He showed me a red GTA. It was beautiful and awesome. However, I needed to see this car in a black color. He then took me to see their GTA (Trans Am), T Tops, in black color. I told the salesman, "This is the one!" He had no idea I was referring to the famous Night Rider on the TV series.

Now, instead of Michael Knight driving this GTA, now it was going to be Michael Torres, driving away with this gorgeous GTA. We sat down to do the numbers. The dealership would take my Chevy as a trade-in. They weren't offering me much for it because of the year and mileage on it, but I took their offer anyway. The difference in monthly payment would be $150.00 out of my pocket (pocket change), plus the bank's monthly share of $300.00 for the next three years. It was definitely a done deal! I customized my GTA to look like the Night Rider on the TV series. I cut the steering wheel on top and bottom and left the middle alone. Now the steering wheel looked like an airplane's steering wheel. I installed lights in the front bumpers that went left and right, just like the Night Rider.

Since the Night Rider talked to Michael Knight, I purchased and installed an alarm system that talked to you! It worked like this. When you set it, if you got too close to the vehicle, it would tell you, "You are too close to the vehicle, step back or the alarm will sound," then it would start counting backwards, "5, 4, 3, 2, 1, EUEUEUEUEEUEUEUEUEU!" My black Night Rider was truly a powerful muscle car. It came with a 350 horsepower engine under the hood. I couldn't resist the pedal to the ground and for that I paid dearly! I received a few speeding tickets. The cops would look at my car while writing me the ticket. They would stare at my cut steering wheel but say nothing about it. I know they had never seen a cut steering wheel before this one!

I once was stopped by a state trooper. He said to me, "I pulled you over because I couldn't see you." Just to let you know, I also tinted my windows too! I felt like saying to this state trooper, "DAAAAAAA," but I refrained from saying anything to piss him off! Now in reality, if you can see me through my tinted windows, then I want my money back, because I would consider a lousy job of tinted windows, don't you agree with me? I continued working for the bank as a computer operator. However, I wanted to advance myself to the next level. The next level would be a choice between staying in operations as a supervisor leading to management or becoming a computer programmer.

In this computer business, you are responsible for your shift's mistakes if you are the supervisor or manager. In the time that I have been a computer operator, I lost count of how many supervisors and managers I have worked for. They all got fired or were forced to resign. As a computer programmer, I knew I would hate it. I knew it was not for me. The money was good but the stress was unbearable! So I chose to stay in operations, where I loved working with computers. However, I was moving on to become a network technician. I became a network technician. My job duties were to install computer terminals throughout our various offices.

After installation of the computer terminal, I would fine tune it before I left it with the user. When I was working at the Network Command Center, I would receive phone calls where I would troubleshoot problems with users using their computer terminals. Some users would call me to reset their user ID or password. When the user tried to log on to their terminals for three times unsuccessfully, their ID got suspended by the system. That was when I got the call. I would reset their ID or password, only after verifying some valid information needed to establish authorizing use of that computer terminal. The best scenario or call from users was when I got a call from them indicating that their terminal would not turn on.

I know this sounds STUPID, but I would ask them, "Is there a chance your terminal is plugged into an electrical outlet on the wall?" I usually got the following response: "Oh, yes, I forgot that I unplugged it to plug my fan." Sometimes it was not a fan; it could be a lamp or a pencil sharpener or something else that needs electricity to run. Another common call would be that their screen was blank and yes, the terminal was plugged into the wall. I would instruct the user to turn the button on the front of their terminal to the on position (turn it to the right). This is the button to save the screen by turning it to the left side. Users do this after they log off their terminals. Many users forget that they turned off the screensaver button when they logged off their terminals.

After working for a few years in Princeton, New Jersey, I turned on the TV to hear the news of the day. I just heard that some explosives were detonated in the basement of the World Trade Center Towers. I immediately remembered my mom, when she was living, warning me about these Towers. She always had a premonition that something very bad was going to happen to the World Trade Centers. Well, the basement was all blown out but the buildings were OK. A year ago, we were introduced to our new vice president, who would be in charge of computer operations. Before his promotion to VP, he was in charge of the computer technical support department, in Princeton, New Jersey.

My boss and I were mentioning our new leader. Sometimes at the workplace, we say things and we assume we are saying them in strict confidentiality. This is sometimes not true at all. So my boss and I were talking about our new VP boss. My boss asked me, "Would you like to work for our new VP?" I said to my boss (in total confidence), "Not really." It seemed that my boss betrayed me and told our new VP what I told him in strict confidence. I guess my boss was doing it to score BROWNIE points with HIS new boss! One month later, I was called into the new VP's office. I thought to myself, "This is odd!" The VP said for me to sit down. He asked me, "How long have you been working for the bank?" Now I knew he already knew this information. I responded, "Twenty-four years this coming June." He then said, "In all these years, you never pursued becoming a supervisor or even a manager?"

So when the VP asked me why I'm not a supervisor or a manager, I responded, "The reason I have lasted 24 years with the bank is for that same reason. Had I elected to become a supervisor or a manager years ago, I wouldn't be here speaking to you, sir!" I then said to him, "I've lost count of how many supervisors, managers, and even VPs I have worked for in these 24 years of service to the bank! All of these bosses have come and gone. Why? All human beings, that we are, we all make mistakes. It seems that it is the policy of computer operations that if a member of the computer staff makes a mistake, the staff member gets fired. Sometimes, depending on the mistake and impact on the business of that day's mistake, the supervisor and sometimes the manager gets fired as well."

In a year's time, I could make as much money as a supervisor or a manager, just by working my overtime hours. All I had to do was do my job and to not make any big mistakes. I'm not saying I have been perfect for 24 years. My mistakes were small and had minimum impact on the bank's business. So to sum it up, the biggest mistake I would have done in my career was to have accepted a promotion to supervisor leading to manager and eventually getting fired. I didn't think the VP liked the fact that, YES, I'd gone through a few VPs in these 24 years of employment with the bank!

He then told me the purpose of me being summoned to his office. He told me that there was going to be a major layoff of computer staff members. He told me that I'd be among those being laid off. So my time at the bank would be in a matter of days, months, or the most, a year left! Sadly, I will not last five more years and retire at 55 years old, with 30 years of service to the bank. This is and now was my ultimate goal! So I told Luz about the meeting with our new VP. If and when I was to get laid off, we would have a problem keeping up with our mortgage payments on our house, which we dearly love.

I called a lawyer friend of mine for legal advice. He said, "Mike, because you are still employed, declaring bankruptcy under Chapter 7 or, for that matter, Chapter 13, is out of the question." He did suggest talking to the bank that held the mortgage and voluntarily giving up the house to them. Selling the home was not an option because we did not know how long I was to be employed. In essence, as he called it, we would be abandoning the house.

Unfortunately, it is what it is. So we took his advice. We rented an old house in a nice Jewish neighborhood in Lakewood, New Jersey. We signed a six-month rental agreement. My idea was to sign a month-to-month agreement, should we need more time before the AX fell down!

I turned 50 years old in May 1995. I also completed my 25th year of service with the bank. On a Monday morning right after I reached this milestone, the phone rang. I was alone. I answered it. It was the VP of computer operations. He never called any of the staff members, unless you were part of management. He very rarely called anybody at home. So when I heard his voice, identifying him, I knew immediately this was not good news at all! He then told me these words in which I'll never forget: "Mike, I got some bad news for you. The bank has decided to downsize the computer staff and therefore thirteen managers, one computer operator, and one network control technician are being laid off today." The network technician was, of course, me.

I was shocked, in spite of the previous meeting with the VP earlier in the year, when I was warned of the layoffs coming down. I don't know why I asked him, perhaps because I was in shock. I asked him, "Do I have to report tonight for work?" He said, "No." He then said, "You will receive a separation package in the mail to sign and mail it back to the bank. The package will include a name and phone number of a contact, should you have any questions." At this time he hung up the phone. I was all alone, at home. I knelt down near the couch and began to pray to my Lord. I said, in my prayers,

"Lord, I ask for your guidance from here on. I do not know what the future will bring, but you do, in Thy name, amen." I later found out my name and residential phone number were immediately deleted from the employees' phone register.

All the friends I made in the twenty-five years of employment with the bank, when they heard the news they wanted to hear from me but had no way of contacting me. I had sufficient grounds to sue the bank on grounds of age discrimination (50), my annual reviews on job performance were above average, and I had an excellent attendance record. The bank obviously knew this, so in the separation papers, a form I had to sign stated that in order for me to receive my 25 years of severance pay, I agreed not to sue the bank. I agreed not to sue the bank simply because I was being put out on the street, now unemployed, and desperately needed my severance money to survive.

Had I not had to sign that form for my severance pay and received my severance pay, I would have definitely sued the bank because it wouldn't have mattered how long it took to hear this case in court. I would have my severance pay to live on financially for a while, as well as my unemployment check. As I said before, I'm a saver of my hard-earned money. I wanted to be set financially when I retired from the bank at age 55 and live on my 401(k) savings, along with the bank's retirement service of 30 years' pension. Well, my goals fell short by five years. When I informed the bank that I wanted my 401(k) money, to my surprise the bank did everything possible to discourage me from withdrawing my 401(k) funds.

The bank brought to my attention that because I was only 50 years old and not 59 and a half years old, I would take a hefty hit from the IRS on taxes, should I withdraw now. How dare they try to convince me to leave my money with them, when they were sending me to the streets! What would happen to my funds, should the bank decide to merge with another firm? Hey, the bank made a lot of money overseeing other company mergers. That department was called Mergers and Acquisitions. What if my previous employer declared itself bankrupt, what would happen to my funds? The bank made me one last offer of monthly payments of $901.00 until the funds ran out.

I refused the offer. The bank then cut me a check for my 401(k) balance and another check for my severance pay for 25 years of employment. I took the checks and yes, I did suffer the taxes the IRS took as well. It is what it is! Now that I was unemployed (first time in 25 years) and 50 years old, what now? What do we do now? Where to? Luz and I discussed our next move. I told Luz that I would like to move back to my island of Vieques, Puerto Rico. I was born on that island. I told her I would try to get a job there, working for the United Postal Service. I would get preference points on the entrance exam, provided I passed the exam with 70 points.

I would be awarded an extra 10 points on my score, thus pushing me up to 80 on my score because I'm a military veteran. Luz flatly said no. Her reasons were that Vieques Island was too isolated. Another reason was that at our age, should we need medical treatments, we would have to go to the mainland for that. What about medical emergencies? These were very valid reasons not to move to Vieques Island. So far, since Luz and I got married, we had moved five times. We started out in the Bronx, New York. We lived in an apartment for our first six months of marriage. We then moved to Staten Island, New York. There we found a lovely duplex apartment that we rented for a year. Our son was born in Staten Island, NY.

We then were offered to move to our babysitter's home, to rent their second floor, with access to three rooms in the attic, should we need it. This became very convenient for everybody's sake. We lived in Staten Island for three years. Then our daughter, Becky, was born. Luz and I decided to purchase our very own home. So we moved to Jackson, New Jersey. Why Jackson, New Jersey? Luz has a sister who lives in Jackson, New Jersey. We visited her several times and we loved the area. That was why we chose to buy a home in Jackson, NJ. We ended buying a two-story Cape Cod home for a reasonable price on my G.I. Bill as a veteran.

Not too long after we moved to Jackson, NJ, my sister-in-law offered us a chihuahua terrier, yellowish tan-colored puppy dog. My son was three years old and my daughter was only three months old. My son loved the puppy and named her Cindy. As the years went by, every night Cindy would climb up the stairs, look at me, and I would tell her who she was sleeping with tonight. It was always either my son or my daughter. Cindy was one of the smartest dogs I had ever owned. Again, not too long after we moved to Jackson, NJ, Six Flags Great Adventure Amusement Park was built in Jackson, NJ.

For the first year the park opened, all residents of Jackson, NJ, were admitted free into the park. We visited the park very frequently that first year, as you can imagine, for free! We lived in Jackson, NJ, for twelve years. So did Cindy, our dog. She passed away and was buried on the other side of our fence, where it's called the common grounds area. My kids and I actually had a burial ceremony for Cindy. I built a large aboveground pool for the kids to swim in. I also built a basketball court for my three-year old son to play basketball (eventually).

As it turned out, the entire neighborhood came to our yard to play basketball. After the years went by, I noticed the walls of the pool were beginning to deteriorate. One day my twelve-year-old niece jumped into the pool from the all-around deck I had built. When she jumped into the water, one of the side walls of the pool opened up and water began to pour out! I told my niece, "You broke my pool and now your mother (my sister) has to pay for it!" Of course I was kidding, but I said it in an angry tone of voice and she got very scared! I did tell my sister, "Now I have an excuse to build an in ground pool!" We lived in Jackson, NJ, for twelve years. While we are on the subject of Jackson, NJ, it was at this time that I was invited to join a new musical group by the name of ECOS MISIONEROS (English translation: Missioners Echoes).

It was a split-level home that we saw for sale and we checked it out and fell in love with it. We sold the Cape Cod home in Jackson and purchased the home in Howell, NJ. This house was located in a cul-de-sac block entrance. We loved it because it was like a private entrance and exit, only for the homeowners living in that particular block. We lived in Howell, New Jersey, for ten years. We loved this house and as much as we loved it, we ended up surrendering it to the mortgage bank, since I couldn't afford making the mortgage payments once I got laid off my job. Once we surrendered the house, we moved to Lakewood, NJ. This was a nearby nice Jewish neighborhood

There we rented a nice but old house. We signed a six-month rental agreement. After the six-month rental agreement, I was going to sign a month-to month agreement until the day came when I would be laid off from my job.

CHAPTER XX

ECOS MISIONEROS

I always have admired musicians. I admire their ability to play a musical instrument, especially when they play it well. As I mentioned before, the church that we transferred to, the pastor's wife played the guitar. She was part of the worship group. I enjoyed the sound of the guitar. I asked the pastor's wife if she could give me lessons. She said yes. I said to her that I would meet her at the church for the lessons. She said not to worry that she would come over to my house for the lessons. I said, "OK."

She began to give me guitar lessons at my house, like she promised. However, as time went by, I kind of lost interest in learning how to play the guitar. Perhaps it was because I later found out that she was only interested in my stepdad. As I mentioned before, he was very handsome, according to the ladies of the church. So I quit the guitar lessons when I was fifteen years old. At the age of sixteen years old, I joined our junior high school band. I chose the trumpet as my selection of a musical instrument. As a junior high school band member, I learned how to read music and play the trumpet. My stepdad purchased my very first trumpet. I decided to better myself in playing the trumpet. I began taking private lessons. A music teacher was recommended to me by my church members. I found out that his music studio was in the Bronx, New York.

This meant that I had to take the subway from Manhattan to the Bronx, in order to arrive at his music studio. I didn't mind the ride because I like riding the subway. I took trumpet lessons once a week. It was at this time that I decided to trade my starter trumpet for a much better upgraded trumpet, which produced a better sound. My stepdad paid for the difference in price. I joined the church worship group, where there were two other trumpet players. I was very happy to become the third trumpet player in the group. When the time came when I moved to Jackson, New Jersey, I began to search for a church to attend and worship. I found a church close to Jackson, NJ. It was located in the next town called Lakewood, NJ.

215

I became a member of the church. In the church were some musicians. The director of the music group found out that I played the trumpet and he invited me to join the worship group. His son played the synthesizer. His brother played the guitar. The director also played the guitar. The director's wife sang, as part of the worship group. After a while, the director decided to form a musical group. He asked me if I would be interested in joining the new musical group. I said yes. I would be delighted to be part of this new musical group. The new musical group was to be named LOS ECOS MISIONEROS (translated: The Missioners Echoes). I have come to realize that my Lord has been with me, all these times, protecting me from all kinds of mishaps, dangerous situations, etc., etc., because it is my DESTINY to serve HIM musically.

When I was the president of the youth church group, my pastor requested me to preach to the youths. It would be my very first time to bring a sermon to the congregation and to the youths. I was attending the Bible Institute at this time. At the Institute, our instructor always said to us that we should always write an outline of what we were going to talk about when asked to preach. So I had my outline with me ready, on the day I was to preach. Well, the unthinkable happened! As I began to get ready to preach, someone decided to turn our giant fan on, which was behind me. As the fan went on, in high speed, so did my outline, flew out the window! So I relied on my LORD to give what I had to preach to the youths on that memorable day! I remember looking at my pastor and he was just smiling from cheek to cheek! I never became a preacher but I certainly became a musician. So now I'm serving my LORD in the music department with this new musical group: The Ecos Misioneros.

The Ecos Misioneros consisted, at first, of eight members:

1) The director, who was the lead singer and played 1st guitar.
2) The director's brother, who sang and played bass guitar.
3) The director's son, who played the synthesizer and sang.
4) The director's wife, who sang.
5) George, who shared in the lead at singing.
6) Tony, who played 1st trumpet.
7) Myself, who played 2nd trumpet.
8) Tito, who played the drums.

These were the original members of The Ecos Misioneros.

As we grew in popularity, we were invited to perform in many churches as well as participate in many concerts. We were often asked if we could bring the sermon to their congregation. We were so blessed in that department as well. For our preachers, we had Oscar, our director's son. We also had Carlos, the director's brother, who also preached.

Our main objective was to bring saved souls to the Lord through our music. Our other goal was to someday record our music in a music studio. It was an expensive process. In order to meet our expenses, we needed to raise funds for recording, transportation, and so forth. Weddings became our main source of revenue. When we were asked to perform in one wedding invitation, that wedding led us to another wedding invitation, and so forth. With the wedding revenues, we were able to purchase a school bus. We eliminated some of the back seats so that we could load all of our musical instruments and equipment to the back of the bus. We painted the bus in a white color on the outside and printed: "The Ecos Misioneros." That was so COOL! Now we were going places.

We were now able to go on tour throughout the States. One of our very first states we went to was Upstate Buffalo, New York. On this trip was when we first obtained our very first flat tire! We received tremendous help from one of the members of the church. I now remember a church we were invited to perform. This church was located in the Bronx, New York. We later found out that while we were performing our music to the congregation, our gasoline was being stolen from the bus. It seemed that the neighborhood we were in was not the best. We found out about the gasoline while we were heading back home on the New Jersey Turnpike, when our bus suddenly stopped. The bus had just run out of gas. A state trooper helped us out by taking me and another member to the nearest gas station to obtain gas for our bus. This was a very cool experience for me because I had never ridden in a state trooper's vehicle before and he just gunned it to the nearest gas station. That was absolutely COOL!

Our dream came true about recording our music. We were able to record our music on three different occasions on three different cassettes. On our third recording, "Mansion Celestial" (translation: "Heavenly Mansion"), my wife, Luz Elba, wrote a song, titled "Plegaria" (translation: "Many Tears"), which we had the honor of recording. In this same recording, I decided to sing with the rest of the singers by joining them in a medley of songs. Santos, who was our studio sound engineer, listened to the track where I'm singing, and he said I sounded like I was out of breath and choking to death. He had no choice but to delete me from the rest of the voices. He said to just stick to my playing the trumpet!

On one of our many concerts, we were invited to accompany a wellknown evangelist at the time to go with him to the Dominican Republic for six days. On this missionary trip, the evangelist instructed us to bring two luggages of clothes and shoes. One luggage would contain our personal items. The other luggage would contain items for donations to the community where we would be staying nearby. When we landed at the Dominican Republic airport, there was a truck waiting for us to pick up our luggage, musical instruments, and our equipment. We got on a bus with the truck and truck driver following us to our destination.

After a four-hour ride, we reached our destination: Barahona, Dominican Republic, which is located on the south side of the Republic. On our first day, we went grocery shopping for the food we were to eat for the next six days there. Among the items suggested we buy were Billy goats. The Dominican locals really know how to cook, especially the goats. We were treated like celebrities. Every morning we had a large fan group made up of beautiful children, waiting for us to come out and meet them. The pastor of the church we were visiting was in need of transportation, which was usually motorcycles. The

Ecos Misioneros came together and pooled our funds and donated him a motorcycle. He didn't know how to thank us!

On our final day, we were asked to perform at the city's stadium. We were honored to perform our music for all of them. When we said our goodbyes, we left our second suitcase to be distributed among the local community. Many of our band members decided to also donate our personal suitcase as well, since we saw how high the poverty level had reached them. We then headed out to Puerto Rico for another six days. We stayed in Barrazas, Carolina, where we performed on our third day by an invitation by the local pastor. We then continued our Puerto Rican tour to the island of Vieques, Puerto Rico, where we performed on our second day, at the invitation of my cousin's church. Church members were so happy to see and hear a live musical group performing at their church!

On our 12th day we finally boarded a flight to Newark, New Jersey. The Ecos Misioneros continued to perform throughout the States, churches, and providing our music at weddings throughout the area. We even got to be incorporated! As all good things eventually end, The Ecos Misioneros was no exception. The musical group began in the 1980s and ended in 1988. You can Google "Ecos Misioneros" and hear our music.

On our first recording, we have Tony playing 1st trumpet. I played 2nd trumpet. On a sad note, Tony has passed on to be with the Lord. However, on our second and third recordings, I did the entire trumpet playing, even though it may sound like a mariachi playing trumpets! The group members, although now dispersed throughout, are still very active indeed. One group member is a world evangelist. Another member is currently a pastor. Another member is in radio broadcasting. The rest of us are very much active, playing our musical instruments, or singing praises to our Lord. As for me, for the past twelve years I've lived in Jackson, New Jersey, and continue to be part of my church's worship group.

Although I will remain in the church I'm now assisting, we have decided to move to Howell Township, not too far from Jackson, New Jersey.

CHAPTER XXI

FLORIDA

Why Florida? Well, Luz has two older sisters already living in Palm Bay, Florida. So we agreed on relocating to Palm Bay, Florida.

I informed my two sisters-in-law about our plans to move to Florida. I needed help in finding a house to rent until I got established there. So I began the process with the help of one of my sisters-in-law. She found a nice house and I sent her the deposit for first month's and last month's rent as required by the realtor in charge of renting out the property. I also sent her some personal information about us, also required in order to do a background check. I signed a six-month lease with an option to continue leasing month to month, if needed. I hired a car transporter to bring my Trans Am to Florida. I also hired a moving company to move all of our household items and bring them to Palm Bay, Florida.

Luz packed everything we owned into boxes of various sizes. The mover then loaded everything into the truck. One of the movers picked up this 30-gallon container that was very heavy. He then asked us what we had in the container that weighed so much. He then answered his own question: "ROCKS?" Luz and I both looked at each other and turned our heads towards him and said, "YES!" These were mineral rocks that Luz collected over the years. She had them spread all over her garden. She didn't want to leave them behind. These rocks are very special. They glitter at nighttime and shine bright during the daytime. Luz and I drove to Florida like so many other times, when we came on vacations to Disney World. However, this time we were not returning back to New Jersey.

We arrived at our rented home and had to wait for the moving truck company to arrive with our household items. However, the truck bringing my Trans Am arrived shortly after we did. The movers' truck arrived a few hours after the Trans Am arrived. The movers would not unload until they were paid for the moving service. I paid them the balance due upon their arrival. The movers began to unload the truck and place the items into the house. To our surprise, the furniture fit perfectly into its designated areas. Luz and I were exhausted after moving. So far, since we got married, we have moved six times. We started out in the Bronx, New York.

Now that we had settled down in our Florida rental home, we could proceed in our next item of our list of things to do. I went to the unemployment office to file for my unemployment benefit checks. I met a very helpful lady who worked at the unemployment office. She heard my case and suggested that I should file for out-of-state status, being that I got laid off, not in Florida but in New Jersey. She said that in my particular case, I would be entitled to more monetary benefits than Floridians. She said that I was entitled to collect up to nine months of benefits. That was somewhat encouraging news for me since I had never collected unemployment benefits ever, in my working years. So all I had to do was call the unemployment office in Albany, NY, and answer the standard questions and my check would be mailed in the next three days. As it turned out, even though I was looking for work, the starting salaries of jobs being offered to me were far less than what I was presently collecting from my unemployment benefit checks.

So I continued looking for work and collecting unemployment checks. As a matter of fact, I ended up collecting unemployment benefit checks for the duration of the entire nine months. When the nine months of collecting unemployment benefits expired, I went to the unemployment office. There I saw and spoke to the same nice lady who took care of me, nine months ago. She informed me that an extension of three more months was just granted to the unemployed. Well, since that definitely included me, I took the extra three months of unemployment benefits.

During this entire time while I was looking for employment and not finding a decent-starting salary job, Luz and I were busy finding a future place to live. We first had to look for a nice lot to build our future home. We were looking for a corner lot. We found one we liked and called the realtor. The owner of the lot lives in Kentucky. We negotiated the price of the lot and the realtor paid the difference in what I was offering and what the owner was asking. The realtor was not about to lose the commission on this deal.

My wife and I wanted to build a custom home, since we shopped around and saw many model homes. Some models we liked their kitchens, other models we liked part of their floor plans, and other models we liked their various styles of bedrooms and bathrooms. So because of these reasons, we decided to hire a custom builder of homes. Luz should have been an interior decorator. She sat down with the custom builder salesman and designed the entire home. In her design, she chose the kitchen of a model home we saw and she made some small modifications. She did the same with the floor plans of other model homes, again with modifications to our taste.

She continued designing the bathrooms, porches, windows, doors, in the same fashion, with slight modifications to our style. During the construction of our home, we were constantly paying a visit to our salesman. Every time we stopped by to see him, our purpose was to upgrade something to our home being built. He loved the visits because to him it just meant more money out of our pockets. On the other hand, the owners of the custom home builders hated us. They hated us because they didn't like changes. Well, we told our salesman, who was very nice, that the owners were in the wrong business of being custom builders of homes.

We were renting a home, within walking distance of where we were building our home. Luz and I became very good at reading blueprints. For instance, one day we paid a visit to our home being constructed and we went inside to our master bedroom. We had been told that this room was completed. We then realized and confirmed it, by looking at the blueprints, that there were no steps leading to the Jacuzzi bathtub. We informed the builders. In another occasion, we went by the house again and we realized the outside windows were too high. We specifically asked for low windows.

We informed the builder and they sent out a crew to reinstall the windows at the correct level that we had specified. To correct this problem, the crew had to break the blocks on the walls and bring them down a bit in order to reinstall the windows. This was just another reason why the owners didn't like us. Hey, it is what it is! The contract we signed with the custom builder was done in the following manner: Four payments to be rendered after four stages of construction of the home were finalized. While the home was being built, we also hired a pool company to build our in ground pool that, by the way, Luz designed with a quarter and a half-dollar.

There were some problems building a pool while at the same time building a home in the same location to each other. Would I have known this, I would have built the pool after the home was completely built. One thing we began to get annoyed at was the construction workers of the house, throwing their trash inside the pool. The pool wasn't completed, so there was no water inside the pool. At the end of all the construction, finally completed after eight months, we moved, again! The cost of our custom home, with all the upgrades, came to a total of $175,000.00. The pool, with all the animal mosaics installed at the bottom of the pool, came to a total of $45,000.00. It was all paid in cash money.

The salesman of the custom home builder wanted to know how much money I brought with me from New Jersey. I said, "I brought two checks with enough money to cover all construction that has been just completed!" I like to save! Now that Luz and I, and my son and daughter, were living in our brand-new custom home and new in ground pool. It was time to concentrate on me getting a decent-paying job. My unemployment benefits ran out after a year. I continued to look for work. One of my neighbors owned a cleaning business. He and his family cleaned offices in a medical building.

I asked him if he could use an extra person to join him in helping with the cleaning. He said, after thinking about it, "As a matter of fact, I just picked up a new account in another medical building, and I can use the extra cleaner." I said, "How much do you pay?" He responded, "$6.00 an hour, all work done in the evenings." In my mind, I said to me, "You got to be kidding me!" One thing I learned from working for a company for 25 years. I often went to work sick because I didn't want to inconvenience my boss in getting somebody to work for me and have to pay twelve hours' overtime. The many sacrifices I made in those 25 years of loyal service to the company, and at the end, the way this company treated me on my dismissal, so unworthy of this treatment! I said to myself, "From now on, I'm going to be loyal to myself and no one else."

So I said to my hiring me, new boss, neighbor, "I'll tell you what, make it $6.25 an hour and I will join your cleaning team." He thought about it for a long time. Mind you, I was only asking for a lousy .25 extra! He finally said, "OK, you got the job." Now, in my mind, I already made up my mind that if another employer out there offered me $6.35 an hour, I was gone, in a heartbeat! This was how it was going to be from now on! While I was working my butt off, cleaning doctors' offices and bathrooms during the evenings, in the daytime I was looking for work to better myself with a better-paying job. I said, "There's got to be something better out there!"

I found a job along my field: electronics. This job was a daytime job. Starting pay was $8.00 an hour. So I gave myself a $1.75 raise just by switching jobs!

It entailed making cell phones inside circuitry. The only drawback I found was that it was an assembly type of work. I'd never worked in this type of environment before. You have to be accurate and work in a fast pace; otherwise, if you are slow, you will hold up the assembly line and the workers will get very mad at you. Again, even though I was working in electronics, I wasn't thrilled working in an assembly line type of work. To be honest, I began to hate my job and there is no fun in that. So I kept looking and hoping for a better, suitable job for me.

Way back when I was living in New Jersey and took my kids to Florida on vacation to Disney World, on one occasion, I suited up and went on an interview with a respectable company in the city of Palm Bay. Their staff was considered to be very professional. I interviewed for a computer operator job opening. To my surprise, the firm hired me. However, I told them that I was living in New Jersey. Well, the interviewer gave me 50 days to begin working for them! Of course, in order to be available in 50 days, I would have to sell my house, and do all the preparations for moving and all that.

So I declined the job. Well, now that I was living in Palm Bay, Florida, I did go back to that firm for another interview. They granted that interview, and yes, I was hired again. The job was the position of a computer operator, working in the evenings, at a starting pay of $10.20 an hour. I do love working with computers, which is down-my-alley kind of work. Since the starting hourly rate was better than the assembly job, which really I hated, although it was working evenings, I accepted the position! So I began working with computers again. I was informed by the firm that hired me that I beat out a few applicants for the job. Of course, I thank the firm for hiring me.

A while back, before my unemployment benefits ran out, I took the United Postal Entry Exam. Even though I passed it, I only scored a 72 and a score of minimum 70 is passing. As a veteran, I was given the opportunity to take the exam more than once. With a score of 72, the USPS will not call anyone for an interview. So I kept taking the exam multiple times in order to get a better score. The second time I took the test, I scored an 80. Even with an 80 score, it is not enough to get an interview. The third time I took the test again, I scored an 89. Even though my score was improving, it was still not enough for an interview to be granted. So I took the exam one more time and this time I scored a 98.

What was so good about getting a 98 was that as a veteran, I was awarded 10 points to my score. So now the USPS had my exam score as 108 and that assured me of an interview with the USPS office! Now, way back when I graduated from high school with a Civil Service Courses background, I always intended to work for the government. So at 19 years old, I took the Entrance USPS Exam. I failed the exam and could not take it again. So I gave up working for the USPS. On the same week I began working for the firm in Palm Bay, as a computer operator, I received a letter from the USPS for an interview on Monday morning.

Since I was working in the evenings for the firm, I was able to go to this interview in the morning. I was instructed to go into the postmaster's office and wait there for him. He took a while to come into his office because he was a very busy boss. So since I was waiting for him, I took the liberty to look at the walls of his office, where I saw many photos of soldiers in Vietnam. When the postmaster entered his office, I was looking at the photos on the walls. It turned out that the postmaster was a ranger in Vietnam. He asked about me. I said, "Sir, I was a medic in Vietnam." He thanked me for taking care of the wounded troops in Vietnam.

We talked about Vietnam and shared our experiences there. I say we spoke about Vietnam for about 20 minutes. He then said, "Let's discuss about the job offer with the USPS." He told me that he expected an honest day's work for an honest pay. I said, "I'll do that and I will not let you down." He then congratulated me on getting one of three jobs being offered by the USPS on that day. This time I beat out a whole lot of applicants competing for these jobs! I was starting at $12.00 an hour, and for Florida, it doesn't get any better than that! Oh, and I was told I had to work on Saturdays. Well, you can't have everything!

I went back to my computer operator job that evening and handed my resignation letter, giving them my two weeks' notice. With the two weeks' notice, it would be a total of three weeks working for the firm. My boss went BERSERK! I told him that I was leaving the firm! He reminded me as to how many applicants they interviewed and refused in order to pick me. After he found out I was going to work for the USPS, he said to me, "When I retire, I better get my check in the mail ON TIME!" Like I said before, I'm only looking out for myself from now on. I began working for the post office after my two weeks' notice given to my other computer operator job. My new hours were 6 A.M. to 2:30 P.M., Mondays, Tuesdays, Thursdays, Fridays, and Saturdays.

I loved the schedule except working on Saturdays. Now that I had my evenings free to myself again, I began to think how I could make use of this free time. Way back when I was collecting my unemployment benefits, I took my grandson to school every morning. Not only that, I even opened up an account with the cafeteria, to have breakfast with him every morning. Well, while I was having breakfast every morning with my grandson, the head custodian asked me, should I ever want a job working for the school district to come and see him. At the time, the position was open to work only four hours daily, during the evenings. I just couldn't survive on working only four hours a day, which would be only 20 hours a week.

But now, I was open to making some extra money. So I paid a visit to the head custodian of the elementary school, and guess what? He remembered me. As it turned out, one of the custodians developed some back problems and she was unable to continue to do her job as a custodian. So the timing was just right for me to apply now for the job. So I accepted the position as her replacement, working Monday through Friday, 6 P.M. to 10 P.M. The hourly rate was $8.00 an hour. I continued working both jobs: the USPS (Federal job) and school (Brevard School Board - State job). However, the school was offering me more hours per week. So I began working six hours a day, Monday through Friday.

My new hours were 4 P.M. to 10 P.M. It was a long day for me but I got used to the new schedule. I later found out that all I had to work for the school system, to become vested, was six years or more. So at my age, this is very important. All I had to do was to continue working for the school board and after just six years, I was eligible to collect a pension from Florida State. It might not be too much money but I'd take whatever they offered when retirement got here. I was also working at the post office and also I expected a pension from the Federal Government, when retirement got here. I also found out that the post office would give me credit for military time served. So for me, that would be two years of postal service credit time.

Now, the post office has a lot of military veterans working for them. Upon me speaking with these veterans, they oriented me on all of unknown claims to the Veterans Administration that I was entitled to. When I was in Vietnam, I broke out with a rash all over my body. I went on sick call and it became documented on my medical records. As it turned out, from time to time, my skin began to itch, and I scratched it, and I developed a rash from it. When I completed my active service with the U.S. Army, I submitted a claim for my rash. The VA awarded me a 10% compensation for service-related cause.

Over the years, ever since Vietnam, I continued to have these nightmares when I went to bed and fell asleep. Many times I kicked, punched, or even screamed, and my wife had to wake me up by slapping my face. I'd wake up with sweat all over my body. My wife suggested I see a VA doctor for my nightmares. At this time I mentioned it to my veteran friends of mine at the post office. They all agreed with my wife and suggested I make an appointment with the VA to see a medical doctor. The veterans at the post office mentioned to me that it seemed to them that I was suffering from something they called PTSD (Post-Traumatic Stress Disorder).

So I made an appointment with the Veterans Administration. They gave me an appointment with the Mental Health Clinic Department to see a psychiatrist. To me, I didn't see the connection with me having nightmares and seeing a psychiatrist. I'm not crazy or felt like I was going crazy! I explained my nightmares to the psychiatrist at the VA Mental Health Department. His diagnostic of my nightmares was, as he said to me, "Chronic Post-Traumatic Stress Disorder"! The disorder had been stemming from my tour of duty in Vietnam. He suggested I continue to see him and he would prescribe some medication to deal with my PTSD.

The veterans at the post office suggested that I submit a claim for PTSD, since I didn't have this condition before I went into the military. After a year, having sessions with my mental health doctor at the VA, my nightmares continued off and on again. Because of the mental health doctor's diagnosis of me having Chronic PTSD, the VA awarded me a 50% service-connected disability. I now have a total of 60% service-connected disability. I then began to hear about Agent Orange that was spread in Vietnam. The Agent Orange was spread throughout Vietnam in order to kill the foliage on the trees and grounds.

Many soldiers coming back from Vietnam began to complain about various skin defects. Well, since I already had a skin condition, I submitted a claim to the VA for Agent Orange affecting me. The VA denied my Agent Orange claim. The VA informed me that Agent Orange was not spread in the area that I was station at. OK, I agreed that Agent Orange was not spread where I was stationed, but what about all the soldiers coming out of outside our area that we treated? Many times we just never thought about wearing gloves or masks. We just concentrated on tending to their wounds. We could have been contaminated at this point. The VA would not hear our argument, so they stood by their decision.

A few years later on, I was diagnosed with diabetes. Since the VA was treating me for diabetes, I submitted a claim for it, since I heard from the veterans at the post office that the VA had linked diabetes with Agent Orange. The VA awarded me 20% service-connected disability for my diabetes. I now had a total of 80% VA service-connected disability. I did my numbers as far as pensions were concerned. I was now lined up to collect a pension from the Florida School Board of Brevard County, Thrift Savings Plan, from the USPS, and Federal Reserve Service from the USPS. I'm also due to collect from the Social Security Administration as soon as I turn 62 years old.

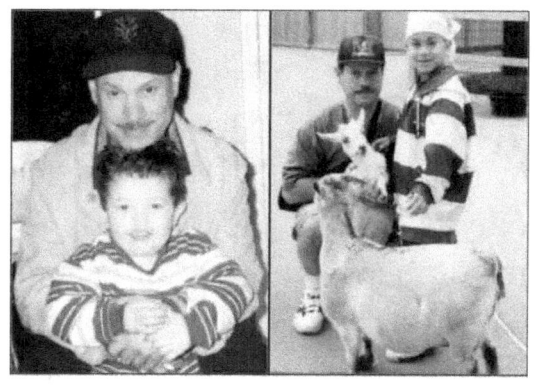

I'm presently collecting from the Veterans Administration a monthly check for my 80% service connected disability. This will give me a total of five pensions to live on for the rest of my life. Luz will be eligible to collect her Social Security, which will be added to my five pensions. So like I said, I did my numbers and I came to the conclusion that at 62 years old, I did not have to continue to work anymore. I began to work at the age of 10 years old and I'm definitely retiring at 62 years old.

On December 28, 2007, I walked into the postmasters' office and informed him that Monday, December 31, 2007, was going to be my last day working for the USPS. I said to him, "Sir, I'm retiring on Monday, last day of this year, 2007." On Monday, December 31, 2007, I walked into the principal's office of the elementary school I was currently working my evening hours and asked to speak to her. I said, "Ma'am, I'm retiring today." She said, "Oh, no, we and staff and the children are going to miss you, Mr. Mike." She then wished me good luck on my retirement and to see the school secretary to submit my retirement paperwork.

I informed the Veterans Administration about leaving the workforce. I told them that I was retiring from both jobs I had. I also informed the VA that my diabetes has progressed from taking pills to adding insulin injections on a daily basis. A few months later, the VA sent me a letter from its regional office in St. Petersburg, Florida. The letter was to inform me that since I was now unemployable, the VA was awarding me with 100% Service-Connected Permanent and Totally Disabled Veteran! I found out that with this new rating from the VA, not only my monthly VA check would increase in the benefit amount, but also I was now eligible for free dental service at the VA Clinic. I also was eligible for a discount on my tags at the motor vehicle.

As for county taxes, I only hav to pay very little, next to nothing. This would be a huge benefit as far as my mortgage monthly payments, I also found out that many local department stores give veterans a discount for their purchases of merchandise. I even found out that certain restaurants do the same discount. I then informed the Social Security Administration that I was filing for my early retirement benefits.

CHAPTER XXII

RETIREMENT

What do I do now that I'm retired at 62 years old? I don't like fishing but I enjoy feeding the fishes. There is no joy for me in hitting a golf ball to a tiny hole I can't see for 300 yards and then go looking for the ball!

Well, when I visited the amusement parks, whether it was in Jackson, New Jersey, or at Disney World in Orlando, Florida, and I observed the clowns, it seemed to me that many were actually not into it. I didn't find them funny at all. So I took up clowning! I had to do something! I began to spread the word around that I was available for birthday parties. My fee would be $125.00 for two hours of clowning around. Well, it wasn't too long that went by that one of the parents of the church I belonged to asked me if I would be willing to attend their little girl's 8th birthday party. I said, "Let me see my schedule." Why, of course, my schedule was empty! I said yes.

We agreed on my fee and I wrote down the date.

My very first GIG as a clown, I couldn't wait! I went out and purchased a clown outfit, along with some mighty big colorful shoes. I also purchased a red round nose, and a few colorful wigs. On the day of the party, I asked my wife, Luz, to do my face, and what a face she did! I told her, "Not bad for the first time doing my face." We couldn't stop laughing! I wanted my clowning to be funny, interesting, and to show my future customers, a one-of-a-kind clown. So, I'd been preparing some magic tricks to incorporate into my show that I had up my sleeves! I was definitely ready for my first birthday party.

I arrived on time, as per instructions, by the parents. However, I could not drive with my giant clown shoes, so I did the driving with my colorful clown socks. The birthday girl was so happy to see me. Then she took me by surprise when she asked me for my name. You know, come to think of it, I completely forgot to give my clown (ME) a name! So I said to her, "You, my little sweetheart, give me a name." Well, she said, "Since you play the trumpet in church (she knew who I was), why don't you call yourself TRUMPO THE CLOWN!" You know, that little eight-year-old sweetheart just gave me my clown name and from that moment on I have been known as "TRUMPO THE CLOWN"!

Like I said, I had a few tricks up my sleeve. For example, I gathered all the children around the table where the cake was. I then took out my coloring book. I then fanned the coloring book from front page to last page. All the pages were blank. I closed the book and said to all the children that we needed to fill the book with something to color it in many different colors. I then fanned the book again from front to end, and now the book was filled with different figures in black and white. I then closed the book again. I then asked the boys and girls, "What are your favorite colors?" Some responded with a popular color like red, others said yellow, others preferred purple, and others liked the color green. I then opened my coloring book, and there, all the colors the children picked now appeared in all the figures in color.

I then closed the book and told the children, "I have to get my coloring book ready for my next showing." I then opened the book from front to back and showed the children that all the pages were blank again! So I then rolled up my left sleeve. At this very same time, one of the younger children said to me, "You are one of us!" You see, since I was in full costume, when I rolled up my left sleeve, the child noticed that I had the same color skin as he had! So at that moment he realized that I was just like him! I continued my performance, after I was pleasantly surprised by the observance of that little marvelous child. I proceeded to take out a quarter out of my giant pocket.

I told the children in front of me that on the third try I was going to make the quarter disappear right through and to inside my left elbow. Not only did I have the attention of the children, now the grownups began to gather with their children. On my first try, I squeezed very hard the quarter into my left elbow with my right hand. After a few seconds of the quarter sticking to my left elbow, it fell to the desk that was in front of me. I then showed my curious audience the red round mark that the quarter just made to my left elbow. I again picked up the quarter with my left hand and transferred the quarter to my right hand. I again squeezed the quarter, very hard into my left elbow, but this time the quarter did not fall down to the table, like the first time.

It stayed stuck in my left elbow. So I took the quarter with my right hand and knocked it down to the table. I then showed my audience how red my left elbow had become where the quarter was about to enter my left elbow. I reminded my audience that this would be the final time for the quarter to go through my left elbow. So I said to the children, "Now children, pay close attention to the quarter." I picked up the quarter, for the last time, with my left hand and transferred the quarter to my right hand. I then proceeded to squeeze as hard as I could, to where the red mark of the quarter in my left elbow was showing. I then slowly slid my right hand off my left elbow, and blew air to the left elbow, where the quarter was supposed to be at. As I slowly removed my right hand from my left elbow, the quarter was not there.

Immediately the children, almost all of them at once, told me to open my left hand. They were all so sure that the quarter was in my left hand. When I opened my left hand, to show all of them, the quarter was not there! They could not believe that the quarter went inside my left elbow! This was all done in front of their eyes! For my next trick, I really had to work on it to perfect it. I took an empty large balloon from my pocket and filled it with air, with my air pump, until it was completely filled. I then proceeded by taking a very large pointed needle and piercing one end of the balloon and continuing to push the needle through the inside of the balloon until it reached the other end without popping the balloon, of course. Then I would retract the needle in the reverse order that it went in until the needle was fully out of the balloon.

I then gave the large needle to the birthday boy or girl, for him or her to stick the needle into the balloon, and of course, the balloon POPPED!

Another trick that I had modified was converting a one-dollar bill into something else. I would drop a dollar bill into a black strainer, shake it, and ask the birthday boy/girl to put their hand into the strainer and take out what they found inside. Well, the birthday person would pull out a two-dollar bill. I found out that most children had never seen a two dollar bill; therefore they called it a fake bill. So I modified the trick into converting the one-dollar bill into a five-dollar bill! The trick was done in the exact manner that I described above.

Of course every magician does card tricks and I'm no exception. For my card trick, I gathered all my children into the middle of where the party was being held. I also invited the parents because these tricks are as intriguing to grownups as they are to the children. Once everyone was in front of me, I asked the children to sit down. I then proceeded with my card trick. I laid out ten different cards on the table, or the floor, as long as everybody could see the ten cards. I then asked the birthday boy/girl to point to the card that he/she was selecting, BUT not until I left the area where they were. I would go to another room. When he/she was done selecting the card, then they could call TRUMPO back to the cards. I had instructed the birthday person to point to the card but never to touch the card.

I came back to join everyone and face the ten cards that I laid out. I then S L O W L Y touched each individual card once. I then began to pick up each card one at a time until I reached the last card. I paused a bit and I proceeded to pick up the last card. When I had the last card in my hand, I turned to the birthday person and I said, "You selected this card!" Of course HE/SHE and everyone around acknowledged that I did pick the right card. This trick was usually requested to be repeated again and again and sometimes again and again! At the end of each party I got parents asking me for my business cards and how much I charged for clowning around birthday parties. Sometimes they wanted to book ahead for a party, so I began carrying a calendar with me.

I once was contacted by the City of Palm Bay, Florida. They were having a fundraiser at the park. The town officials wanted TRUMPO THE CLOWN to entertain the children that were attending with their parents. I remember reading the contract, which stated that in the event the fundraiser was cancelled due to inclement weather, I would not be paid! Go figure! I'm enjoying my retirement. Luz and I are now going on different cruises and seeing other countries. So far we have visited Belize, Honduran, Bahamas, and now we are planning a trip to Europe. We are also driving to see other states. We just drove to Tennessee and absolutely loved the scenery with its mountains.

In this trip to Tennessee, we were traveling in an RV that my son-in-law rented. On a Friday evening we decided to attend to a comedy club in downtown Tennessee. We were fortunate to sit in the second row of the front seats. At one point the MC asked the audience for a volunteer. My two grandsons and my daughter HOLLERED real loud, "HERE!" They were all pointing their fingers towards me, of course. The MC turned towards us and said to me, "Sir, could you please join us here at the stage?" I didn't have any choice but to do so. The MC then asked for two more volunteers to also join him at the stage. The MC had three empty chairs for us to sit.

For some reason, I ended up sitting in the middle chair, with the other two volunteers sitting at each end of me. The MC asked each of us to introduce ourselves to the audience. The MC also wanted each of us to inform the audience what we did for a living. The gentleman on my left said his name. Then the MC asked him, "What do you do for a living?" He said, "I'M the CEO of Hair CLUB FOR MEN."

There was just one thing wrong with this statement. He was COMPLETELY BALD! I then ran my right hand through my BALD head and made a funny face, like TIM CONWAY would make. In fact, some people in the audience said that I did look like TIM CONWAY, as well as DON RICKLES.

I then noticed that the audience began to laugh hysterically at me when I did that. That was just the beginning of my COMEDY DEBUT! The MC immediately noticed right away the audience reaction towards me. He then asked me to identify myself. I did. He then asked me what I did for a living. I said, "I'm a retired IBM technician." The MC then said to the audience, "Did you just say retarded computer technician?" I then said to the MC, "I'm sorry for my accent but I did not say retarded, I said, retired computer tech." The MC said, "SURE." I don't think the MC believed me but why would I lie? The MC then asked the gentleman on my right to introduce him. He told us his name and that he was a manager of a firm.

The MC seemed to believe him. The MC was standing behind us. We probably looked like three STUPID IDIOTS on stage, facing the audience, not knowing what to expect next. The MC took out a long, round, spongy noodle-type of a toy (maybe). He was holding it in his right hand. He then gave us some instructions to follow. He told us that when he hit our heads with this long round noodle, we were to open and shut our mouth repeatedly. Do this as many times as he hit our heads with the noodle. This was to be done once the music began to play. However, when he hit our shoulders with this noodle, we switched from opening and closing our mouth to clapping our hands over and over, as many times as he hit our shoulders.

When the music began to play, it was some country melody, the MC hit the guy to my left, on his head, and he began to open and shut his mouth repeatedly. The MC then switched and began to hit the guy on my right, on his shoulders, and he began to clap his hands, over and over. Then the MC turned towards me and began to hit my head with his long, round noodle. I began to clap and clap my hands. The guy on my right said to me, "You're supposed to open and close your mouth, not clap your hands!" So I began to open and close my mouth but I kept clapping my hands at the same time. I looked at the guy on my right and made a face and pushed my shoulders up and down, showing him that I was very CONFUSED!

He then told me to leave my shoulders alone. I later found out that the audience was laughing at me so much, they thought I was part of the comedy club staff. Like I said before, I have a resemblance of looking like Tim Conway and Don Rickles at the same time. My daughter, who was in the audience watching me, had to run to the bathroom!

Throughout this entire routine in which I was only supposed to clap my hands or open my mouth repeatedly, depending on how the MC hit you, I completely got it all wrong, from the very beginning. I broke all the rules. At one point, I opened my shirt and began to dance while I was still sitting down in the chair.

The audience went berserk, laughing at me. I guess with the music getting faster and louder, I lost it after it was all over and the music stop, the MC said, "Folks, it doesn't get any better than this!" We walked off the stage to a standing ovation and that my friends were my Comedian DEBUT! This is now one less item on my bucket list before I kick the bucket! I will continue to travel with my son-in-law, who loves to travel to various places along with my daughter and my two grandsons. Even though my son-in-law is quite away from his retirement, he takes time off to travel.

I enjoy the company of all my five grandsons. Why just today, Super Bowl Sunday, I was playing Basketball with my two youngest grandsons. I enjoy going to their basketball games. When Baseball season is here, I attend those games, too. When my oldest grandson was playing AAU (Traveling Baseball league), I did a lot of traveling to see him play Baseball throughout the States. When my next to the oldest grandson was in the school swimming team, I enjoyed watching him swim against his competitors.

I only have one granddaughter who is ten states away from us. And I wish she was closer to us so that my wife and I could do things together. However, when Luz and I get together with her, we take her shopping!

I'm also enjoying my five year old great grandson. Let me just say that from the very beginning, as he began to walk, he also began to climb. He climbs everything! I will not be surprised if when he gets older, he decides to climb Mount Everest. When we babysit him, which is three days a week, he begins by climbing my back, then continues to my shoulders, and finally climbs to my head, where I give him a victory flag to wave!

While on the subject of flags, many veterans have flags on their tombs. I hope to have mine someday too. However, it crosses my mind of what I will be doing when I leave this earth. It is said that Elvis Presley died in his bathroom of his Graceland, Memphis, Tennessee, home. As for Nelson

Rockefeller, it is said that he died of a heart attack, caused by a Bang. As for me, my best way to go will be during my beauty sleep, hopefully.

As I close this final Chapter and finish this book, which was also on my Bucket List, before I kick the bucket, I bid you farewell my friends and may my Lord bless you always!

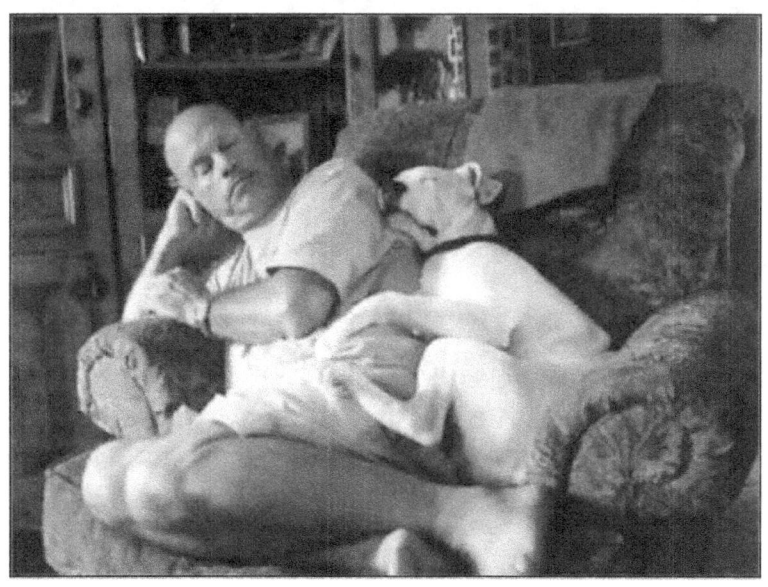

Juggernaut and I taking our customary NAP

SUMMARY

WHY I NAMED THIS BOOK DESTINY

1. I was informed that because I was born on the day before Mother's Day, the doctor slapped me!
2. At age six I stepped on a broken bottom of a bottle.
3. At age nine I touched a cigarette lighter in my uncle's car while the lighter was on.
4. At age ten I was roller skating in the streets of Philadelphia and my wheels got in between the trolley car rails and I fell down. A car behind me put on his brakes. I ended up under the motor of the car. The driver pulled me out of his car with not a scratch on me.
5. At age twelve I was bitten in the back of my neck by a horse.
6. At age twelve I was thrown by a horse. (Not even Superman was ever the same after that.)
7. At age twelve I missed my turn, getting hit in the head by a pick.

8. At age twelve I refused an invitation by my cousin to go fishing. The boat sank and my cousin swam two hours to reach the shore.
9. At age twelve I was robbed while playing handball at Jefferson Park, East Harlem, New York City.
10. At age thirteen I was shining shoes at 42nd Street and Times Square and the guy whose shoes I was shining, invited me to his hotel room. I declined the offer.
11. At age thirteen a police officer chased me out of 42nd Street for shining shoes.
12. At age nineteen I was rear ended in a car accident.
13. At age twenty-one I spent 365 days in Vietnam; 58,220 American soldiers lost their lives.
14. I jumped the Staten Island Ferry after it just left the dock.
15. While boarding the Vieques Island ferry boat, I slipped and one of my legs went in between the moving ferry and the dock.
16. I was working on the 7th floor in a building on Wall Street, NYC, when a fire broke out and the 7th, 8th, and 9th floors were consumed by fire.
17. I worked at the World Trade Center.

18. I've fallen off several ladders. The last fall was from the top of a seven-foot ladder where I landed on my butt, hitting the back of my head. I average one fall per year.
19. I've had several low-sugar convulsions where my mind goes blank.
20. I have a couple of kidney stones stuck in my kidneys.
21. I have an enlarged prostate that I'm trying to shrink.
22. I had my left eye turned sideways but it has been corrected.
23. I've touched several live wires several times without being electrocuted.
24. I was bitten by a dog once and the dog died!
25. I lit a Roman candle rocket and it was facing me, burning my shirt and arms.
26. I had a firecracker explode up near my ear. I temporarily lost my hearing.
27. While riding my English Racer once I put on the brakes and I flipped over, landing on my back.
28. While my buddy and I were bringing a refrigerator down from the second floor, the refrigerator rolled down the stairs with me under it.
29. I fell down the stairs from the second floor, holding my infant son in abassinet.

30. A Vietnamese barber, before giving me a haircut, twisted my neck, I heard it crack but not broken.
31. I have been surrounded by neighborhood gangs several times and survived them.
32. My winter coat was put on fire by a street gang while I was wearing it.
33. While walking on Vieques Island beach one day, a coconut fell from a30-foot palm tree, missing inches from hitting my head.
34. My two STUPID buddies of mine and myself decided one day to go on a search-and-find mission through the jungles of Vietnam in civilian clothes. We found a leprosarium colony in the mountains. On our way back we met a platoon of Korean soldiers. Had we met the enemy instead of the Koreans, I assure you I wouldn't be writing about it.
35. I continue to have nightmares with my stepmother beating me up.

www.ingramcontent.com/pod-product-compliance
Lightning Source LLC
Chambersburg PA
CBHW060906120626
46553CB00001B/230